Making It Personal

Making It Personal

Algorithmic Personalization, Identity, and Everyday Life

TANYA KANT

OXFORD
UNIVERSITY PRESS

OXFORD
UNIVERSITY PRESS

Oxford University Press is a department of the University of Oxford. It furthers the University's objective of excellence in research, scholarship, and education by publishing worldwide. Oxford is a registered trade mark of Oxford University Press in the UK and certain other countries.

Published in the United States of America by Oxford University Press
198 Madison Avenue, New York, NY 10016, United States of America.

Library of Congress Cataloging-in-Publication Data
Names: Kant, Tanya (Lecturer in media and cultural studies), author.
Title: Making it personal : algorithmic personalization, identity, and everyday life / Tanya Kant.
Description: New York, NY : Oxford University Press, [2020] |
Includes bibliographical references and index.
Identifiers: LCCN 2019033684 (print) | LCCN 2019033685 (ebook) |
ISBN 9780190905088 (hbk) | ISBN 9780190905095 (pbk) |
ISBN 9780190905118 (epub) | ISBN 9780190905125 (online) | ISBN 9780190905101 (epdf)
Subjects: LCSH: Human-computer interaction.
Classification: LCC QA76.9.H85 K36 2020 (print) | LCC QA76.9.H85 (ebook) |
DDC 004.01/9—dc23
LC record available at https://lccn.loc.gov/2019033684
LC ebook record available at https://lccn.loc.gov/2019033685

CONTENTS

LIST OF FIGURES

ACKNOWLEDGMENTS

I would first and foremost like to extend a heartfelt thanks to all those who participated in this project. Without your time and contributions, this book would not have been possible.

So many people have contributed to the writing of this book, which began life as a doctoral project funded by the UK Arts and Humanities Research Council. An enormous thank you to my former supervisors and colleagues Caroline Bassett and Sharif Mowlabocus for their support, thorough and insightful feedback, and at times seemingly prophetic advice—your guidance has been invaluable in steering this research in fruitful directions.

I am grateful, too, to all the members of Sussex University's Media, Film, and Music faculty who have helped me along the way—especially Michael Bull, who offered warm yet practical guidance as I stepped into my first adventure in academic book publishing. I am especially indebted to my fellow early career researchers and friends who helped with the development of this work from a PhD thesis to book: Lizzie Reed, Ryan Burns, Rachel Tavernor, Rachel Wood, Gemma Cobb, Ben Litherland, and Russell Glasson, and to the ECR group more widely. Thank you also to all of those colleagues and friends who are too numerous to name but who have offered a friendly ear and steadfast support; I feel very lucky to have been able to complete my first monograph in an environment that fosters just the right mix of kindness, hard work, and occasional but much-needed comic relief.

I am extremely grateful to the editorial team at Oxford University Press for their work in developing the book, especially editor Sarah Humphreville; thank you for your time and patience.

Last but not least, thank you to my Mum, Dad, and sisters Danielle, Lisa and Kerry for their love and understanding through what have been some difficult times together, especially over the years it has taken to develop this work. Thank you, as always, for your support, steady stream of hugs, and for generally putting up with me.

1

Introduction

Making it Personal

This site uses cookies for analytics, personalized content and ads. By continuing to browse this site, you agree to this use.

—Microsoft (2018)

As News Feed evolves, we'll continue building easy-to-use and powerful tools to give you the most personalized experience.

—Mosseri, Facebook Newsroom (2016)

Previously, we only offered Personalized Search for signed-in users, and only when they had Web History enabled on their Google Accounts. What we're doing today is expanding Personalized Search so that we can provide it to signed-out users as well.

—Horling and Kulick, Google Blog (2009)

Hints that a user's web experience might now be "personalized" exist in innumerable traces all over the web. The preceding quotes from Microsoft, Facebook, and Google highlight that dominant web platforms are increasingly embracing what Fan and Poole call the "intuitive but also slippery" practice of personalization (2006, 183): the

Making It Personal. Tanya Kant, Oxford University Press (2020). © Oxford University Press.
DOI: 10.1093/oso/9780190905088.001.0001

process of delivering apparently individually tailored online content and services to users, based on their habits, preferences, and identity markers. Like countless other "cookie notices"[1] across the internet, Microsoft's notification mobilizes the term to explain why users' data are tracked and managed across its platforms (MSN, 2018; Microsoft, 2018; Live, 2017). Facebook's Newsroom Blog reinforces the platform's long-standing aim to provide a "personalized experience" for each of the site's two billion active monthly users. Invoking the sentiment that "Personalized Search" can *only* be convenient for its users, in 2009 the Google Blog stated that Google Search would provide personalized search results to all users, even to those who were signed out of their Google account. Since then, Google and its parent company Alphabet have increasingly incorporated personalization into their other services: Google Maps, Google's mobile app, and YouTube all provide extensive forms of personally recommended content as well as individually tailored information, location, product, and entertainment suggestions. Though only implied in the preceding quotes, the information collected in order to "personalize" web user engagements is of course also used to target those same users for monetization purposes (Bassett, 2008; Jordan, 2015; Jarrett, 2014; Cheney-Lippold, 2017).

Beyond their references to personalization, the previously quoted sentiments of Microsoft, Google, and Facebook share three similarities worth highlighting. First, they treat personalization as "intuitive"—it does not need to be explained or justified, but is instead presented as a practice that simply exists, apparently for the indisputable advantage of the user. Second, they exemplify that personalization is indeed "slippery"—it can be mobilized in various contexts and can be used in ways that do not necessarily reveal the specificities of *what* is being personalized (simply "content," "experiences," or "search"), or *how* or *when*. Third, they highlight that in the context of the contemporary web, the user does not enact personalization—it is implemented and controlled by the system, platform, or service. You do not personalize your web experience, your news feeds, your playlists, or your product recommendations. Rather, with the help of a multitude of personal data collected by platforms as you go about your day, your needs and interests can be algorithmically inferred and your experience "conveniently"—and computationally— personalized on your behalf.

It is the "slippery" but certainly not "intuitive" presence of personalization in my own web experiences that provides the foundational motivations for this book. When I first embarked on this research in 2013, evidence that some components of my web experience were being "personalized" took the form of web advertisements for recommended products I had recently browsed on retailers' web sites. At the time, these ads seemed crude, invasive, and usually entirely ineffective. Sometimes I would be served advertisements for products that I had literally just bought, rendering targeting me based on my previous browsing habits so precise as to be absurdly pointless. At other times, the knowledge (produced by the ad's very existence) that the advertisement could only be generated by algorithmically "watching" me triggered a feeling of privacy invasion that worked to overshadow any relevance the ad might have to my personal preferences. The enduring presence of these tailored advertisements, delivered across platforms and in a variety of different formats, serves as a reminder that the personalizing of my daily web trajectory inherently involves relinquishing some form of personal data in exchange for the free content and services I access and enjoy on a daily basis.

The delivery of targeted advertisements continues to persist, yet is now accompanied by other personalization practices that seemingly transcend but, as I will argue, never fully depart from targeted marketing. It is no longer only clearly bounded "recommended ads" that are personalized, but also content, services, interfaces, and indeed simply "experience" itself. "Personalization" is frequently cited as the reason for the collection of user data on some of the world's most visited sites. For example, on their privacy policy, owners of AOL and Yahoo Verizon Media state that they use the data they collect "to provide you with personalized experiences and advertising across the devices you use" (Oath, 2019). Elsewhere on the web, entertainment web site BuzzFeed states that they allow third parties to place cookies on their sites "to serve ads to you and to target you with tailored (or personalized) ads" (BuzzFeed, 2018). Online music player Spotify's Privacy Policy explains that "we need to understand your listening habits" through data collection "so we can deliver an exceptional and personalized service specifically for you" (Spotify, 2018). The term is used to validate the tracking of users in hundreds, if not thousands, of user-facing cookie notices, including the Microsoft notices cited earlier, but also on sites such as eBay, BBC, *The Washington Post*, Huffpost, Twitter, *The Financial Times*, Instagram, Imgur, and innumerable others. In fact, on the contemporary web, personalization seems both undefined and just "there": in their Privacy Policies, eight out of the ten most popular global web sites cite the "personalization" or "customizing" of users' experience as one of the primary reasons for harvesting users' personal data.[2] Though "personalizing" and "customizing" are distinct terms, on these sites they are used to mean the same thing—to "know" a user's individual tastes, preferences, identity components, habits, or desires in order to deliver some content, interface, or service that is deemed to be individually suited to that user.

The concept that our web experience can and should be personalized is similarly reflected in marketing of the "intelligent personal assistants" (Myers et al., 2007) increasingly finding a place in homes and on mobile phones. Google's Assistant, Apple's Siri, and Amazon's Echo claim to be able to computationally manage the demands and routines of daily life by delivering customized content feeds and recommendations in the form of news bulletins, task scheduling, traffic updates, geolocative weather information, product notifications, or recommended playlists. The "personal" touch of these digital assistants is enacted by algorithmic mechanisms and is framed as technology that can preempt users' informational needs, what texts they would like to watch, listen to, or consume, and what products most suit their preferences. As well as algorithmically anticipating a user's desires, the development of voice-recognition means that users can now *command* these assistants: for example, Alexa users can instruct the device to play the song of their choice, or indeed do their homework for them (TwitterClips, 2018). As I will explore in this book, these personal assistants paradoxically promise to make the best choices *for* us, while simultaneously offering to action the choices *we* make, thus leading to the everyday yet complex entanglement of human and non-human agency that I argue throws the sovereignty of selfhood into question. Of course, like the web services cited earlier, the makers of these personal assistants reserve the right to collect user data as a means of personalization—and monetization.

As the number of personalization technologies grows, so too has the data-tracking practices used to infer users' everyday habits and sociocultural economic practices. Contemporary data-tracking strategies include harvesting users' browsing histories, Facebook "likes," purchase histories, search histories, geolocation, app interactions, the photos they upload, their mobile and other audibly detectible conversations, the comments they write, their home appliance use, their cross-device activity and IP (internet protocol) addresses, the content of their texts and emails, their commute to work, their "friend" connections, their "moods," their song downloads, their credit history, their movie/TV viewing choices, and their gaming high-scores, among a host of other traceable everyday actions. Contemporary data-collection devices include Amazon's Echo Look, which takes photos of its users so that "algorithms curate your closet for you, organizing your looks by season, weather, occasion and more" (Amazon, 2018), and Facebook's Messenger app, which according to its Terms of Service can "record audio with the microphone *at any time and without your confirmation*" (Facebook Messenger App, cited in Watson, 2013, my emphasis). The wording of such claims omits the specifics of exactly how and when these data will be used (indeed if at all). However, it is implied that upon being collected, these snippets of everyday life are collated and connected to other user data sets in order to construct and manage behavioral profiles, user demographics, and other configurations of user identity.

Taken together, the preceding data-collection mechanisms are deployed as part of a now common socioeconomic exchange imposed by platform providers and accepted by web users. Users are routinely expected to submit to the tracking of their everyday habits and movements in exchange for free, convenient, and personalized services. Though framed as a kind of additional benefit by the platforms that implement it, personalization is not primarily the "goodwill" gesture it might first appear: instead, the user data relinquished as part of the exchange are *the* driving economic resource of the contemporary free-to-use web. As a market model, this data-for-services exchange seems to be functioning very successfully: in the first quarter of 2018, the world's biggest data tracker, Facebook, made $11.97 billion in revenue (Protalinski, 2018), in large part generated via personalization systems that deliver "micro-targeted advertising" to its users. The tracking of users is a widely accepted form of monetization that is legitimized through discourses that imply that personalization results in a "better" web infrastructure for both user and platform. As I will emphasize throughout this book, it is not an overstatement to propose that the drive to personalize web user experience underpins the online economy as we know it.

Profit generation from personalization exists in spite of the privacy scandal that engulfed Facebook in 2018, which uncovered the (mis) use of Facebook user data by analytics company Cambridge Analytica to build "a system that could target US voters with personalised political advertisements" based on extensive behavioral and demographic profiles (Greenfield, 2018). The profiling and targeting undertaken by Cambridge Analytica transpired to be widespread, with the company claiming to work on campaigns not only in the United States but in the United Kingdom, Cyprus, Nigeria, India, Italy, and over twenty-five other countries (Ghoshal, 2018), sparking debates that personalized political advertising could unduly influence elections and in doing so undermine democracy itself. The scandal threatened to devalue Facebook's share prices, as well as trigger much stricter data-tracking regulations both on and off Facebook (though such threats have as yet had little long-term impact for the platform's popularity). Beyond Facebook, in the same year, the EU General Data Protect Regulation (GDPR) legislation replaced the 1998 Data Protection Act: a move intended to update privacy regulations and give EU citizens more security over how their personal data are used. Though a legal necessity only for EU web sites, the GDPR has had a wider global effect: for streamlining purposes, many non-EU platform providers have complied with EU requirements. From these privacy directives to Facebook's CEO Mark Zuckerberg's recent testimony in US congressional hearings, data tracking has come to underpin the popular web as a kind of necessary evil: condemned as privacy-invading and a potential threat to democracy, yet increasingly legitimized as making possible the apparently essential personalization of marketing, content, and services upon which platforms rely. It is the enduring and widespread implementation of personalization as a market practice that provides one of the many motivations for this book.

It is not only commercial dominance of data-driven personalization that has sparked my investigations: it is platforms' attempts to "know," anticipate, and, as I will explore, *act on* the person implied in the term "personalization." As scholars such as Gillespie (2014) and Bucher (2016) note, though academic scholarship has shown interest in the commercial tracking and anticipation of web users through largely theoretical and quantitative approaches, there has been less research to qualitatively focus on how web users themselves engage with and negotiate the algorithms that seek to "know" them through data. As Bucher states:

> While media and communication scholars have started to take notice of algorithms . . . little is yet known about the ways in which users know and perceive that algorithms are part of their "media life." (2016, 31)

Cohn has relatedly argued that theoretical critiques of algorithms in social life tend to "be overly dystopian or utopian in tone" (2019, 10) in ways that pay little attention to the nuances that go into encountering personalization as part of everyday engagement. It seems then that, as Gillespie puts it, "it is easy to theorize, but substantially more difficult to document, how users may shift their worldviews to accommodate the underlying logics and implicit presumptions of the algorithms they use regularly" (2014, 187).

This proposed disconnect brings me to this book's primary aim: to bridge the gap between "theory" and "documentation" in regard to how algorithmic personalization intersects with users' notions of identity, autonomy, and everyday life. To do this, the book employs a mixed methodology of qualitative investigation, political economy, and critical and historical analysis to understand how personalization has come to dominate the digital landscape, as well as asking *users themselves* how they experience platforms' attempts to track their personal preferences, anticipate their actions, and personalize their experience accordingly. In combining these methodological approaches, it becomes possible to recognize algorithmic personalization not only as an everyday, situated encounter, but also as a macrocosmic "force relation" (Bucher, 2016) that plays a pivotal role in twenty-first-century capitalist web economies. As the individual testimonies featured in the book highlight, when approached from the perspective of web users themselves, data tracking in the name of personalization occupies a routine yet tense place in many web users' everyday experiences. Described by those who encounter it as ubiquitous, convenient, infuriating, unexceptional, persistent, unnerving, and opaque, personalization appears to be an inescapable fixture of the digital landscape, whether welcomed, resisted, or negotiated as somewhere in between.

BEYOND PRIVACY: ALGORITHMIC ANTICIPATION

For many years scholarly debates on the collection of web user data have largely coalesced around issues of privacy—defending it on the grounds of human rights (Lyon, 2014; McStay 2017), relinquishing it in order to use web services (Bassett, 2013; Jordan, 2015; Turow et al., 2015; Peacock, 2014), or indeed questioning the very possibility of it in the context of tracking practices so complex, vague, and opaque as to be in some ways always unknowable (Brunton and Nissenbaum, 2015; Mai, 2016). Since Edward Snowden revealed in 2013 that commercial platforms such as Google and Facebook have been aiding the state "dataveillance" of millions of web users in the United States and the United Kingdom (BBC, 2016; *Guardian*, 2013; Lyon, 2014; Seeman, 2015), popular and academic debates surrounding online privacy have proliferated. Such privacy debates gained further traction in 2018 following the fallout of the aforementioned Cambridge Analytica scandal, with concerns about privacy framed as data "misuse" by data researchers, as compared to the apparently "appropriate" use of these data for commercial personalization. Despite increased public scrutiny, however, there is evidence that user *action* in resisting data tracking remains muted—increasingly, users are aware that they are being targeted, and though many users feel uncomfortable being tracked (Ofcom, 2019), they are resigned to the fact that if they want access to the services they use on a daily basis, then they must sacrifice their data to do so (Turow et al., 2015).

These privacy concerns inform elements of this book, and they certainly highlight the wide-reaching political implications that emerge from the daily surveillance of web users. However, in recent years there has been an increasing critical insistence that "privacy is not the only politically relevant concern" (Gillespie, 2014, 173) when it comes to data tracking. This is because data tracking does not exist, in and of itself, simply to surveil or track users, but to *anticipate* them (Gillespie, 2014; Cheney-Lippold, 2017; Hearn, 2017), to "know" some facet(s) of a user's identity in order to make "personally" relevant some component of experience on their behalf. Crucially, this process of commercial anticipation is executed not with the central aim of "watching" the individual user, but instead to *act* on, with, or against their experience of the web.

Take, for example, Facebook's data-tracking practices. According to Facebook's user-facing page "Your ad preferences" (Facebook, 2018), Facebook has categorized my online behavior into a dizzyingly extensive list of around 390 "interests," which include categories like "Feminism," "Dogs," "Media Studies," "Digital Technologies," "*Orange Is the New Black*," and "University of Sussex." These listed categories do indeed reflect some of my sociocultural, political, and economic interests: others do not ("Quilt," "Whitewater"), some are almost laughably specific or existentially vague ("Clydesdale Horse," "Failure"), and others still remain a mystery as to which of my preferences they signify ("Hour," "Emotion," "Boy," "Sequence," and "Li Ke"). All of these categories exist despite the fact that I have actively resisted explicitly inputting any categories of interest on Facebook: these "preferences" have instead been inferred from my daily trajectory on Facebook and on the web more generally. The inferred categories are used to anticipate the kind of advertising and other content that I might find "relevant," and my News Feed is filtered and adjusted accordingly (Facebook Help Center, 2018). Then, through a process of recursion (Jordan, 2015),[3] this algorithmically adjusted News Feed informs the kind of interests that Facebook might infer from my activity henceforth—the cycle both changes and continues ceaselessly.

Though the name "Your ad preferences" suggests that these categories of interest only affect the advertisements I see, in actuality these "advertisements" usually take the form of "sponsored posts," which are not formatted as ads, but rather are integrated into Facebook's News Feed as news and entertainment content. This blurring of advertising and "organic" content (Hardy, 2015) highlights that though personalization processes often revolve around targeted advertising, its effects extend far beyond marketing. Most importantly, this cycle of inferring my interests and amending my News Feed highlights that the very purpose of collecting data about my daily trajectories is to act accordingly on my experiences—to algorithmically make decisions for me, to filter, reorder, or display content or information in such a way as to be personally "relevant."

It is through examples such as "Your ad preferences" that the role of algorithms in the online personalization process becomes explicit, and it is the power afforded to algorithms to determine what is "personal" *on behalf* of the user that makes this form of personalization distinct from other forms of personalized production. After all, practically anything can be individually customized—products, gifts, clothes, shoes, and furnishings can be tailored to suit personal preferences (Getting Personal, 2017; Prezzy Box, 2017; Your Design, 2017); health care (Roche, 2017), social care (NHS, 2017), and educational packages (Personalizing Education, 2017) can be personalized to appeal to specific people. It is, however, *algorithmic personalization* that takes the focus of this book—the *computational* tracking and anticipation of users' preferences, movements, and identity categorizations in order to algorithmically intervene in users' daily experiences. By specifying this form of personalization as "algorithmic," I do not mean to suggest that algorithms somehow exist beyond the human; as scholars such as Finn (2017), Law (1991), and (Oudshoorn et al.,

2004) argue, computational systems are always in some way informed by and situated in very "human" aspects of design, culture, and sociality. Yet, as I will explore in this book, it is algorithms' power to intervene in everyday action—independently from the direct decision-making processes of marketers, platform providers, designers, or users—that imbues them with material and productive power as non-human actors (Latour, 2005).

Though many so-called adaptive algorithms—algorithms that have the capacity to independently intervene in decision chains, such as those used in global stock market calculations—have autonomous and agential powers, the accounts of the web users interviewed for this book highlight that commercial personalization algorithms have particular resonance as decision-makers in everyday web experience. This is due to not only their ubiquity, but also, as I will detail, the *epistemic uncertainties* and *struggles for autonomy* that they create for the users who encounter personalization. Differently put, *who* decides what is "made personal," how this is decided, and the ways in which these decisions reshape the parameters of engagement have a multitude of implications for the people who encounter personalization —both "online" "and offline." In fact, the ubiquity of networked technologies, combined with the increasing integration of data-mining systems in devices from mobile phones to cars to toasters and even tampons (Hern and Mahdawi, 2018) mean that it is difficult and indeed unproductive to differentiate between everyday "online" and "offline" interactions in late capitalist digital cultures. I therefore join scholars such as Berry (2014) and Sauter (2013) in asserting that the "online/offline" dichotomy is becoming increasingly unhelpful. The contemporary drive to personalize web user experiences is extending rapidly beyond solely online information, and as such this book is based on the premise that algorithmic personalization may be implemented through networked technologies, but is better considered as part of users' "media life" (Deuze, 2011).

By assigning web users the role of "data providers" (Van Dijck, 2009, 47) while algorithmic protocols are assigned as the gatekeepers for deciding personal relevancy, algorithmic personalization creates tense relations between user and computational system that other forms of personalization do not. The accounts contained in this book suggest that it is not only privacy invasion that explains the discomfort some web users feel when confronted with, for instance, a personalized advertisement that follows them around the web. Rather, it is the fact that an algorithm has been afforded the autonomous power to reorder and repurpose users' web experiences in the users' stead. As I will argue, by intervening in and reshaping users' everyday experiences, algorithmic personalization imbues the system with the power to co-constitute users' experience, identity, and selfhood in a performative sense (Butler, 1988, 1990, 1993). Chapter 3 will detail the performative possibilities of algorithms in the everyday—possibilities which, I argue, open up critical considerations concerning the "entangled state of agencies" of user and system (Barad, 2007, 23). This "symbiotic agency" (Neff and Nagy, 2016) opens up new avenues for investigation in understanding the self: as algorithmically *in*formed but user *per*formed, as preexisting the algorithm but also brought into existence by it, and as governed but autonomous.

In recent years an increasing variety of literature has identified the sociocultural implications of web personalization. Most prominently, Pariser (2011) has argued that because web personalization relies on mining users' preexisting browsing and search histories to determine what users see next, personalization reductively reaffirms our existing worldviews, causing "filter bubbles" of consumption. By algorithmically implementing users' future web experiences on their past preferences, Pariser argues that filter bubbles create "you-loops" in identity that only ever reinforce rather than challenge or diversify users' existing sociocultural beliefs. Since the publication of Pariser's work in 2011, filter bubble theory has been further refined, theorized, and debated, and a plethora of research has emerged that has sought to find evidence of the filter bubble at work. For instance, Hosanagar et al. propose in their study of iTunes that "[p]ersonalization appears to be a tool for helping users widen their interests, which in turn creates commonality with others" (2013, 1). Skrubbeltrang et al. (2017) analyze user perspectives of personalization on Instagram, finding that user-led counter-narratives resist the assumption that algorithmic personalization is beneficial to users. Koutra et al.'s (2014) study regarding algorithmically filtered news consumption supports Pariser's arguments, concluding that "people use the web to largely access agreeable information" that ultimately provides a "myopic view" (2014, 8). The contrasting outcomes of these studies suggest that algorithmic personalization creates different outcomes for users depending on the context. They also highlight the importance of context-specific research in understanding the nuances of theory when applied to everyday web engagement.

The filter bubble is perhaps to date the most widely interrogated theory connected to personalization, leading to guidance and software designed to help users combat its proposed reductive effects (MIT, 2013; Ponsot, 2017). However, the computational anticipation of users' needs, preferences, and desires has also triggered debates regarding the sociocultural implications of personalization not for users' worldviews, but for their own *identities*. Cohn (2019), for example, offers a detailed and comprehensive analysis of contemporary recommendations systems that, as I explore in Chapter 2, constitute one of the key forms of algorithmic personalization that web users most commonly encounter. Cohn argues that the automated, seemingly individualized recommendations offered by platforms such as Google, Facebook, and Netflix work to "shape the contemporary self" (2019, 51) by encouraging and indeed coercing users to "choose" content that is easily monetizable and standardizing, and that reinforces existing dominant frameworks of inequality and neoliberalism. He argues that

> [r]ecommendation systems privilege the "free choice" of users as a synecdoche of their unique individuality, self-worth and authenticity, while, in fact, always guiding the user towards certain choices over others in order to encourage them to better fit with those the system recognizes as being like them. (2019, 7)

For Cohn, personalized recommendation systems thus have the *potential* to reduce and reconfigure the self in ways that suit the logic of neoliberal capitalism. I stress *potential* here, however, because for Cohn, the ideological controls imposed by recommendations are never total. He argues instead that users find ways to "critique, ignore, laugh at, negotiate with, and otherwise respond to recommendations" (2019, 8) in ways that suggest user resistance and subversion to algorithmic recommendation as a mode of sociocultural governance.

In his work on data profiling, Cheney-Lippold argues that data trackers' attempts to profile users into "emergent categories" such as "high cost" or "celebrity" (2017, 4), or "terrorist," "male," or "female," as well as other niche categories, work to produce the self purely in and through computational mechanisms. For Cheney-Lippold, the shifting, modulatory, and performative productions of algorithmic classification mean that our identities are not simply reflected in data—instead, increasingly "we *are* data" (2017, my emphasis). Hearn proposes that ubiquitous data tracking has replaced late twentieth-century notions of the neoliberal, self-promoting individual, instead giving rise to the "anticipatory, speculative self" (2017, 74) that is in some ways more fluid than the neoliberal self and yet forever in need of "verification" by big data processors such as Twitter. Bucher has argued that users are becoming increasingly familiar with living through and in algorithmic infrastructures, creating norms in our everyday "media life" wherein we may have come to "see and identify ourselves through the 'eyes' of the algorithm" (2016, 34–35). Other scholars such as Skeggs (2017), O'Neil (2016), and Noble (2018) interrogate the divisive and discriminatory practices inherent in being profiled in and through data; as I will detail in Chapters 2 and 3, the behavioral profiling upon which personalization relies seeks to identify not only web users who function as "valued" consumers, but also those who are de- or undervalued as economic risk. As with the socioeconomic classifications that have functioned to govern and discipline social subjects for hundreds of years, under algorithmic personalization, not all persons are identified and constituted as equal.

Increasingly, then, the algorithmic anticipation of users' identities foundational to delivering "personalized" experiences is being critically defined as a process of subjectivity constitution: this process creates "algorithmic identities" (Cheney-Lippold, 2017, 5), "data doubles" (Lyon, 2014, 6), "database subjects" (Jarrett, 2014, 27), and "algorithmic selves" (Pasquale,

2015, 1) that are designed to *intersect* and *interact* with the identities that they are intended to mirror, represent, and/or constitute. It is these algorithmic configurations that take the interest of this book. More specifically, it is the intersection of these datafied "selves" with the people who are constituted "in" and yet continue to live "outside" of the algorithm that I argue give rise to fresh considerations of autonomy, identity, and the digital everyday.

PERSONALIZATION, POLITICAL ECONOMY, AND THE EVERYDAY

Given its market dominance, understanding algorithmic personalization as a fixture of the everyday demands attention through two methodological avenues: first, how the *principle* of personalization is put into techno-economic *practice*; and second, how the profitable, datafied "person" can be theorized, historically contextualized, and understood as a lived subject position. As such, this book is underpinned by a mixed methodological approach that combines critical political economy, qualitative analysis, and historical and theoretical critique in order to evaluate algorithmic personalization at both microcosmic and macrocosmic levels.

A critical political economy approach is useful for considering how the principle of personalization is put into techno-economic practice because as methodology it is interested in interrogating the role that commerce plays in structuring the social fabric of everyday life. As Murdock and Golding (2005), Mattelart and Mattelart (1998), Greenstein and Esterhuyan (2006), and Bettig (1996) emphasize, political economy is useful for identifying the "underlying social relations" between market logics and individual needs and interactions (Greenstein and Esterhuyan, 2006, 15). Internet scholars such as Fuchs (2011), Bodle (2015), and Jarrett (2014a) highlight that critical political economy helps to clarify and situate the negotiations of web users as value-generating social subjects, always engrained in *but not* always determined by broader sociocultural and economic capitalist contexts.

Though critical political economy is useful for understanding how algorithmic personalization might function as a dominant market principle, it is less helpful for getting to the minutiae of how the people entangled in personalization systems negotiate their status as data providers. Theoretical and historical analysis is useful here to understand what is meant by the "person" at the heart of the personalization process. As I explore throughout this book, though the concept of the "person" might seem self-evident, it is historically and contextually specific formation that has changed over time and continues to be (re)constituted in critical and popular landscapes. In evaluating the ways in which the "person" has come to be understood, it becomes possible to locate the performative power of algorithms in identity production. However, as I will argue and as others such as Cohn (2019) similarly suggest, theorizations of algorithmic governance must also do justice to the ways in which web users exercise their agential capacities in techno-capitalist cultures that might impose forms of discipline on users but do not necessarily determine their everyday engagements.

There is an increasing wealth of literature that deliberates the promises and problems that emerge in algorithmic anticipation—critiques that underpin and inform the findings presented in this book. However, as Gillespie's (2014) aforementioned claim highlights, there remains something of a disconnect between "theory" and "documentation" in regard to the ways in which algorithmic encounters are felt and experienced at the level of the everyday. Though, as the qualitative researchers I reference in this book emphasize, "the everyday" is a complex term that can and should not be reduced to a generalized set of practices or frameworks, it seems important to confront personalization algorithms as part of the fabric of digital life in global late-capitalist cultures. After all, personalization algorithms are used in some way by countless commercial web platforms, and millions of web users frequently produce mundane, habitual, and individualized encounters with a range of advertising, service, and content formats. The same "personalized" advertisement will no doubt be shown to thousands of people deemed to be "interested" in whatever it is selling, the same recommended story or media artifact delivered to a unified demographic of people, and yet each engagement is embedded in a specific time and space, a specific interface, and a specific set of entanglements for the users who confront that (im)personalized text. A series of questions arise for me in these simultaneously personal and impersonal encounters: How can user engagements with personalization be investigated in their specificities? What new insights—and unexpected avenues—of critique might be found in the situated, nuanced, and context-specific accounts of the users who confront personalization algorithms? Are users' encounters really as oppressive or subversive as contemporary scholarship suggests, or are there other ways to articulate and assess the networked negotiations of the people who are both served and (through the monetization of their personal data) serve algorithmic personalization systems? To answer these questions, I argue that we must—by employing qualitative methodologies—turn to the specific testimonies of web users themselves.

BRIDGING THE GAP: GETTING AT LIVED EXPERIENCE

For many decades, qualitative research methods have been employed as a methodology for "getting at" the daily lives of individuals, in order to untangle the complex ways in which meaning is constructed and maintained in apparently mundane, everyday contexts (Warren, 2001; McNeill and Chapman 2005; Maynard, 1994). Qualitative approaches to gathering research data—such as interviews, ethnographies, participant observations, and focus groups—are understood as fruitful, productive, and reflexive avenues to exploring the kinds of meaning-making that occur in daily life. Just as importantly, as the long-established theorizations of Geertz ([1973] 1993) remind us, it is in their ability to capture *rich* and *deep* data that qualitative research methods accommodate critical analyses of social interactions without reducing or condensing their inherent complexity and nuance. As Braun and Clarke state, in-depth and rich data can "record the messiness of real life" in ways that enable critical researchers to put "an organising framework around it" (2013, 20). I argue, then, that it is in exploring the lived experience of the individuals entangled within algorithmic personalization technologies that it becomes possible to critique and reassess broader structural questions in regard to networked knowledge production, autonomy, and selfhood.

The thirty-six individuals interviewed for this research have all, through their engagement with Ghostery, Facebook apps, or the Google mobile app, encountered algorithmic personalization in some form. I offer more details of their recruitment in corresponding chapters dedicated to each study; here however I want to comment broadly on the sampling and analyses methodologies chosen for this book. As Emmel notes "there are no guidelines, tests or adequacy, or power calculation to establish sample size in qualitative research" (2013, 9) and in fact Emmel is amongst a number of scholars who have pushed back against the idea that there is a "right" size of sample for qualitative research. Unlike quantitative sampling methods, qualitative samples can be considered as "invariably small" (Emmel, 2013, 5) because they are rich; full of utterances and modes of expression that take time to unpack and untangle. After all, the point of qualitative research is very rarely for that sample to be "representative" of a larger group, because to "represent" even the smallest of populations assumes that those individuals in that group have a measurable set of qualities that bring together their specific experiences. I cannot stress enough that the same holds true for this research: the accounts of the thirty-six individual interviewed for this project do not and cannot represent the engagement of *all*—or even some—of the web users who encounter algorithmic personalization. As Emmel straightforwardly puts it, with qualitative research "the concern in designing these studies is not how many, but what for" (2013, 3). Their richness and deepness therefore "allow for the interpretation and explaining of social processes" (2013, 5) in ways that quantitative data cannot—because of its very "bigness"—accommodate. Participant samples for this research project were settled on the basis that they were big enough to allow for diversity of responses, but small enough to do justice to respondents' complex, nuanced, playful, glib—and as I will explore consistently tense—articulations on algorithmic personalization systems.

All participants had English as a first or second language and were largely based in the UK (even if temporarily, as with some of the international students interviewed for the Google mobile app project), as well as the US, Canada, the Netherlands, and France, and so this book does not include analyses drawn from anyone who lives outside these dominant language and nation-state boundaries. I call at this point for future research in other linguistic and national contexts, not to "speak for" those languages and countries (in the same way in which my analysis does not "speak for" the UK's, US's or France's web users) but to further expand a field primarily dominated by Anglo-centric studies. As algorithmic personalization practices continue to develop, the opportunities to pursue new research avenues that extend beyond Anglo-centric contexts are becoming more and more apparent. This book explores participants' accounts that are always already located within socio-cultural normative tastes and practices—practices that are specific to their lived experiences and are therefore contingent on context-specific parameters of taste, class, and cultural preference (amongst other factors). In doing so there are a number of key conceptual, political and theoretical claims I want to make, especially in regards to how algorithmic personalization intervenes in and indeed constructs users' sense of self.

As feminist approaches to qualitative methods have emphasized (Cotterill, 1992; Maynard, 1994; Stanley & Wise, 1990; Skeggs, 2004; Haraway, 1988), analyzing everyday experience is a complex task in which researchers much respectfully acknowledge the validity of the experiences reported by research participants while also accepting that the notion of experience should not be taken as unproblematic (Maynard, 1994, 15). "To begin with there is no such thing as 'raw experience'": rather, as Maynard notes, "the very act of speaking about experience is to culturally and discursively constitute it" (1994, 15). Engaging with the "lived experiences" of individuals is always an act of also *constructing* that experience, of framing participants' accounts through a framework informed by the researcher's own research goals, cultural context, and gendered, classed, and raced (among other sociocultural identity markers) subjectivity. This book follows these feminist-informed principles: I look to treat participants' lived experiences of algorithmic personalization as a reflexive, co-constructed dialogue between researcher and participant—while contextualizing these experiences within a wider critical and political economic framework.

Haraway points to the epistemic richness to be found in "situated knowledges"—located ways of knowing that do not rely on or reinforce universal objectivities, but instead take seriously "situated and embodied" epistemologies (1988, 583). In and through situated subjectivity, knowledge can be productively constructed not as universal, common, or unified, but as always contextual, complex, interpretative, and political. However, I join Skeggs (2004) in being mindful not to fetishize subjectivity as a means to inarguably "authentic" knowledge—that is, assuming subjective experience has a kind of possessive ownership of what we can take to be true about the world. In grounding my findings not in quantitative analysis or theoretical hypothesis, but rather in the testimonies of web users themselves, I look to resist generalizing my findings through what would be ultimately "unlocatable, and so irresponsible, knowledge claims" (Haraway, 1988, 583).

It is important to emphasize that as a researcher, I myself am a situated subject. The interviews generated from this project should not and cannot be considered a "blank slate" upon which facts about algorithmic personalization appear: my own situatedness as a Media and Cultural Studies theorist has, like all research, underpinned and shaped the outcomes of this project. Therefore the quest to find meaning in participant encounters must be reflexively—or to use Barad's terminology "diffractively" (2007)[4]—addressed as something created and shaped not only by the participant, but also by the researcher. Ultimately it seems important to note, as Kvale and Brinkman do, that "the research interview is not a conversation between equal partners, because the researcher defines and controls the situation" (2009, 6). Thus, I hope to do justice to participants' accounts while contextualizing their responses within my own critical framework.

It seems pertinent to also briefly acknowledge that the relationship between participant and research is not universally applicable to all interviewees and across all scenarios; a huge number of factors can affect the research-interviewee dynamic, and Cotterill (1992) states that such changeable dynamics mean the researcher does not necessarily always enjoy more authority than the participant during an interview. To draw on my own research experiences, many of the Ghostery interviewees' technical knowledge of data tracking far outstripped my own, highlighting that the researcher is not always more knowledgeable on the research subject than the participant. Conversely, during interviews for the Google app study I felt very aware of the authority bestowed on me as a researcher; in this case the participants were first-year undergraduate students who (understandably given that I had delivered a lecture to them as part of my call for research participants) recognized me as a tutor, and therefore looked to me to "teach" and "lead" them. I explore this further in the subsequent chapters, but here I want to emphasize that qualitative research is filled with considerations such as these, which should be at least acknowledged and accounted for, even if they cannot be completely resolved.

Interview responses were reviewed and re-reviewed a number of times to identify themes and strands that emerged from and in response to interview questions (themes such as user control over data, self-expression, knowledge of personalization practices). I was eager to allow for a range of different responses, but also acknowledge responses that did not necessarily fit recurrent themes across interviews, and yet still merited critical attention. I am cautious to speak of a "saturation point" at which this book's interviewees stopped bringing new themes into our conversations—to look for such a saturation point once again implies a positivist standpoint that qualitative data are about getting "enough" to work with. My research instead looked to find codes and themes that could speak back to current work on algorithmic personalization in ways that open up new opportunities for critical analysis.

As user anticipation becomes more ubiquitous, the ways in which the self is constructed is increasingly entangled with the algorithm. I argue it is *only* through analysis of lived experience that we can fully understand the complex and nuanced ways in which the self is co-constructed by *both* user and algorithm. I stress this in Chapter 3, where I argue that qualitative analysis provides a useful counterpoint to purely theoretical critiques that tend to overstate the disciplinary power of algorithms. The political resonance of this claim is most pressing when users understand themselves to "be" what the algorithm says they are. For example, I argue that Google essentially tells its users they are white, male, and middle-class, which in turn produces a re-negotiation for users' own sense of self, as I expand on in Chapter 6. Crucially, I argue that the political interventions of algorithms into the everyday must be critically approached not just to consider what algorithms do or don't do—or indeed, as Finn poses, what they might "want" (2017)—but how users negotiate and respond to these algorithmic interventions. As the accounts of the participants in this book suggest, though algorithmic personalization does indeed constitute users' everyday encounters, its power is *co*-constitutive: users' identity is constructed by both algorithm *and* users themselves.

BOOK STRUCTURE AND OVERARCHING THEMES

Chapter 2, "The Drive to Personalize," employs critical political economy to map the historical and technological development of personalization as a market-driven principle put into algorithmic practice. It first offers an overview of contemporary personalization systems, paying particular attention to the co-constitutional relationship created between "user" and "system" in the commercial quest to "know" user intention. The chapter then works outward toward algorithmic personalization as a matter of anticipating the user for monetization purposes. Finally, I argue that even as algorithmic personalization claims to aid the individual in the fight against "infoglut" (Andrejevic, 2013), the autonomous decision-making capacities of algorithmic personalization actually create a *struggle for autonomy* between user and system. It is this struggle for autonomy that opens up questions regarding why web users might contest or resist personalization in ways that go beyond the framing of data tracking as a matter of privacy invasion. Instead, this struggle reconstitutes that which can be considered as "the person" at the heart of the drive to personalize the web.

In Chapter 3, "Me, Myself, and the Algorithm," I offer an analysis of how user identities have been understood historically and discursively. This chapter situates the "person" supposedly key to the personalization process within wider critical frameworks that interrogate the role and value of the self in late-capitalist contexts. I consider how the self is performatively constituted both inside and outside of algorithmic governance, arguing that it is only by understanding identity as algorithmically managed and anticipated, but politically "felt" by users as social subjects, that the nuances of algorithmic personalization can be understood.

The subsequent three chapters constitute the qualitative body of this book. Drawing on the accounts of twelve interview participants, Chapter 4, "Hiding Your 'Scuzzy Bits,'" focuses on the privacy tool Ghostery and the ways in which Ghostery users negotiate their positions as (unwilling) data providers in relation to algorithmic personalization. The chapter explores a number of themes that emerged from interviews, which were semi-structured using Ghostery's own marketing tagline and rhetorical sum: "Knowledge + Control = Privacy" (Ghostery, 2014). These themes include the data-for-services exchange (reluctantly) undertaken by participants, as well as the *epistemic anxieties* created by what can only ever be partial knowledge of data tracking, articulated through participant statements such as "Ghostery gives me a false sense of security" (Claire, UK, 2013). I argue that this epistemic anxiety actually *increases* in accounts of participants who could be considered "power users" (Sundar and Marathe, 2010), who framed the use of privacy tools not as a means of meaningful resistance but as an " 'up yours' gesture" (Gyrogearsloose, UK, 2013). The chapter also analyzes how "personalization" fits into Ghostery's rhetorical sum. I explore the disconnect between participants' negotiations with data tracking, which they wholeheartedly resisted, compared to their negotiations with personalization practices, which some welcomed. Finally, the chapter evaluates participants' sense of privacy in relation to critical notions of selfhood. I argue that participants framed their use of Ghostery as part of the desire to "hide your scuzzy bits" (Chris, UK, 2014)—to protect a preexistent, inner, and possessive selfhood which must be sheltered from the dehumanizing threat of data tracking. I propose that this framing of selfhood corresponds to the self as disciplinable state citizen, rather than the self as expressive commercial consumer.

Chapter 5, "Autoposting the Self into Existence," moves away from personalizing processes that threaten an inner self to algorithmic personalization that has the power to (re)write the self. To do so, the chapter focuses on Facebook's autoposting apps—third-party apps such as Spotify, Candy Crush, and Map My Run that are connected to Facebook and are capable of posting status updates on the user's behalf, without the user's immediate knowledge or explicit consent. Though Facebook claims these apps help users "express who they are through all the things they do" (Facebook, 2014), I argue that if users' lived experience is considered, then autoposting takes on a performative power to bring the self into existence. Drawing on the accounts of sixteen Facebook users, the chapter explores how Facebook seeks to algorithmically personalize personhood itself. The chapter focuses on moments when autoposting apps have caused identity disruption or slippage in participants' self-performances on Facebook and beyond—from Spotify posting "embarrassing" song preferences to participants' "invisible" Facebook audiences to Instagram disclosing participants' sexual preferences. I also explore participants' framing of other users' game posts as "chavvy," arguing that such framing reveals complex class dynamics at work that go beyond the established "digital divide." I argue instead that such engagements produce discourses that frame user value not through cultural capital but *algorithmic capital*, wherein social subjects deploy value judgments aimed at other users' orientation toward personalization algorithms. I conclude the chapter by proposing that autoposting thus takes on startling performative and disciplinary power to write the self into existence—a formation of the self that exists alongside but in tension with the anticipated user as inner, possessive, and preexistent.

Chapter 6, "Validating the Self through Google," explores the "predictive powers" (Android, 2012) of the Google mobile app, a personalization technology that claims to "give users the information you need throughout your day *before you even ask*" (Google, 2014, my emphasis). Using a co-observant approach, users were interviewed five times over the space of six weeks. The participants were six first-year, first-term students enrolled in a Media Studies degree at a UK university, recruited largely because they did not explicitly self-identify as preexisting Google app users, unlike the Facebook and Ghostery users examined in the previous chapters. The chapter is structured around the overarching sense of *epistemic trust* that participants invested in the app—despite the fact that the app's personalization capabilities repeatedly failed to live up to participants' high expectations. The chapter explores this tension between faith and failure, and argues that the Google app's personalization framework is in fact deeply apersonal: Google constructs an idea of what "life should look like" that assumes a homogenous and normative subject position. The chapter explores respondents' extensive efforts to "get something" from the app, wherein participants worked hard to orient their everyday practices toward Google in order to be validated as a "doing subject" by the app. I conclude by first considering participants' status as media studies students, and second by interrogating participants' claims that they will continue using the app, "just because it's there now" (Rachel, UK, 2014). I argue that even though participants ultimately failed to find a use for the app, its predictive promise becomes the reason in and of itself to trust in Google.

The Conclusion pulls together the core themes of the book, examining especially the selfhoods that are made—and made possible—under "regimes of anticipation" (Adams et al., cited in Hearn, 2017, 73) that do not necessarily seek to discipline the subject, and yet entangle the agential capacities of user and system in everyday life. I address some future directions for algorithmic personalization, interrogating Google's discursive erasure of the term "personalization" from many of their user-facing marketing materials: what was once "personalized search" is now simply "search." This in no way means the algorithmic anticipation of users is disappearing—rather, it is becoming naturalized as *the only* way to structure and experience web engagement. The conclusion proposes that under the complex yet "black boxed" ways that algorithms seek to know us, "understanding" the algorithm becomes less important than negotiating the interventions of algorithmic personalization into everyday life.

I argue that as user anticipation becomes more ubiquitous, we must consider data providers not through neoliberal discourses that position data subjects as "agents of their own success" (Ringrose and Walkderine, 2008, 228) but instead as *algorithmic tacticians* who deploy algorithmic capital to "make do" with computational anticipation in sometimes playful, usually productive, and always meaningful ways. The ways in which the self is constructed is increasingly entangled with the algorithm—and it is through lived experience that the sociocultural nuances, implications, and interventions become most apparent.

My investigations produce some context-specific findings that evidence the nuances of living life in, with, and through algorithmic anticipation. These findings can be considered as emergent from situated experiences with particular forms of personalization: however, such particularities both underpin and illuminate some core themes that run throughout the book. The following analyses tease apart the ways in which users must confront the uncertainties of how, when, and why they are being anticipated by algorithmic personalization—confrontations that result in *epistemic trust* or *anxiety* in the algorithm, depending on users' acceptance of their positions as data providers. Building on this, another theme emerges: the constant negotiation of the *data-for-services exchange* that web users navigate on a daily basis, even in the act of resisting data tracking itself. Furthermore, the *struggle for autonomy* that algorithmic personalization produces arises again and again, wherein users' agential capacities—outsourced to the system in the name of personal convenience—emerge as a site of struggle between decision-making algorithms and users as autonomous social subjects. These findings are brought together by my claims that algorithmic personalization has the performative power to intervene in, co-constitute, and even bring into existence the self in ways that reveal the deeply impersonal and dividuating mechanics that lay behind contemporary neoliberal rhetorics of "personal relevance."

As I will further explore, the investigative threads of this research come together to interrogate the ways that users negotiate different models of selfhood: models that not just differ from one another, *but that are in tension with one another*. The participant negotiations featured in this book suggest that algorithmic personalization demands that user identities must be constituted as unitary, inner, and fixable, as endlessly expressive, recursively reworkable, and flexible, *and* as identities that can be legitimized by both algorithm and user. Others, such as Szulc (2018) and Jordan (2015), have come to similar conclusions, noting that social media logics create fixed, "anchored" (Szulc, 2018) selves as well as "abundant" (Szulc, 2018) and "recursive" (Jordan, 2015) ones. I contribute to these debates by arguing that these selfhoods are not just structured in culture and algorithms, but are *felt, embraced, reconciled, and rejected* by those users who navigate and negotiate such structures, at the level of the individual—and it is web users' self-identified positions in the co-constitution of these models of selfhood that in part explain these tensions. As the following chapters demonstrate, these tensions must be contextualized within broader historical and theoretical concepts of the self if we are to understand how the anticipated individual has come to exist—and must be analyzed through lived experience if we are to get at the nuances and complexities of algorithmic interventions into everyday life.

2

The Drive to Personalize

The notion that users should be presented with a personalized web experience is now commonplace on many web platforms: from the algorithmically organized interfaces of Facebook, Twitter, and YouTube to digital personal assistants Siri, Echo, and Alexa, there exist a multitude of algorithmic actors apparently ready to fulfill users' inferred needs and preferences. These systems of inference are largely built on the assumption that understanding a user's identity is crucial to the personalization process: algorithmic personalization technologies must, through data aggregation, "know" the user they are seeking to anticipate if they are to infer that individual's wants and desires. This process is cloaked in discourses designed to reassure users that, despite the use of personal identifiers to fix web users to particular profiles, their data are protected, secured, and "anonymized" at various points in data management and brokerage chains. I will unpick the complexity of algorithmically identifying the (in)dividual throughout this chapter, but it is important from the offset to establish that despite platforms' assurances of individual "anonymity," the very point of data tracking is to in some

Making It Personal. Tanya Kant, Oxford University Press (2020). © Oxford University Press.
DOI: 10.1093/oso/9780190905088.001.0001

way identify and anticipate "the user" so as to intervene in the user's lived digital experience.

As scholars such as Bassett (2013, 2008) and Willson and Leaver (2015) highlight, the monetization of user data is reliant on a social contract between user and platform: users must relinquish their data in exchange for the platform's service. Though there are alternative web monetization models gaining more traction—for example, micropayment systems offered by sites such as Patreon or subscription models like Spotify's pay-monthly streaming service—the data-for-services model currently functions as a primary form of revenue generation for the majority of the world's popular online services. More than this though, commercial platform providers frame this contemporary socioeconomic context as inevitable: data in exchange for a (personalized) service is presented as the best, and often the *only*, option for platform users. This begs some questions: How did the ubiquitous tracking and commodification of web user interactions come to dominate as a market strategy? Why is it that other economic models, though not unheard of, have not enjoyed the same success? Indeed, why is it that the once "free" and "public" service ethos championed as the underpinning rationale of the web is now a "free" but very-much "private" service ethos? As I will argue in this chapter, the market dominance of individual data tracking takes on particular—and somewhat paradoxical—significance when historical developments of the world wide web and internet are taken into account: namely, that in the mid- to late 1990s, when the web was still in its infancy, "cyberspace" was largely celebrated for allowing users the freedom of *anonymity*. How then did the shift from the anonymous (or as I will explore at least self-identified) user, to the algorithmically identifiable and anticipated "person," come about? This chapter considers these questions, by defining some overarching characteristics of personalization and charting a history of commercial data tracking. As I will argue, it is by historically mapping both the technological and discursive development of data tracking that it becomes possible to critically understand why algorithmic personalization both seeks to "dividuate" the user (Cheney-Lippold, 2017) but also co-constitutes the individual outside of the algorithm. I also evaluate the development of discourses that champion algorithmic personalization as "aiding" users in their decision-making, even as the same practices also bring about a struggle for autonomy between user and system.

CURRENT PRACTICES OF PERSONALIZATION

Before offering a historical overview, it is first useful to consider what lies behind the term "personalization" when it is used by popular web platforms. The *principle* of personalizing an object, engagement, or interface is relatively simple: as stated in Chapter 1, the term is commonly used to describe imbuing an object, service, or content with increased "personal relevance." However, the term increasingly applies to the *algorithmic* personalizing of a service, informational feed, or experience. When algorithms are introduced to the personalization process, it is not the user who is responsible for the implementation of personalization, but algorithmic protocols that are designed to automatically render an object personal in the user's stead. Notable adopters of algorithmic personalization include Amazon, whose personalized product recommendations work by inferring a customer's product preferences from "what products we look at, when and for how long we look at them" (Mayer-Schönberger and Ramge, 2018, 77). Netflix also relies heavily on personalization technologies to deliver to users movie and TV recommendations—so much so that in 2009 the company offered a $1 million prize to any data science team that could improve their personalized recommendations.[1] In 2017, the platform moved from a user-driven ratings system to an entirely algorithmically calculated percentage system,[2] meaning that at the level of the interface at least, a user's experience of Netflix is structured primarily around algorithmically personalized recommendations. What was once a "5-star" film recommendation based at least partly on popular and collective relevance is now wholly individualized: on Netflix you are now a "97% match" with the film and TV show being recommended. YouTube has similarly increased the prominence of personalized content on their site: whereas in 2013 their "Suggested for You" sections took up a small portion of the Homepage, the user is currently confronted with almost only personalized viewing suggestions upon visiting the platform. Finally, Facebook's News Feed looks to provide users with "the stories that matter most to them" (Facebook Help Centre, 2018) by algorithmically attaching a "relevancy score" to all the possible posts a single Facebook user might see. Inferred through what are called "proxy" actions, such as previous "likes," comments, click-throughs, popularity of posts, and previous connections to "friends," Facebook includes what it algorithmically deems to be the most "personally relevant" posts in a user's News Feed (Mosseri, 2016). As explored in Chapter 1, this form of "organic" personalization is combined with monetizable "suggested posts"—paid advertising by brands and businesses that is targeted to individual Facebook users via user identity profiles compiled of over 52,000 unique data signals (Skeggs, 2017).

In the field of Human-Computer Interaction (HCI), there is a healthy body of work that helps to explain the personalization technologies currently incorporated into commercial use. Encompassed most broadly under the term "recommender systems," common manifestations of algorithmic personalization include "content-based recommenders" that suggest items of relevance to users based on the properties of that item (Said & Bellogín, 2018). There are "session-based recommendations" (found in software such as online music players) that are designed to suggest personalized content within a short and specific time-based period of user engagement (Jannach et al., 2017). As the overarching term suggests, models such as these focus on the *content* of what is being personalized, rather than the user, to infer items of relevance, and though these short-term forms of personalization are becoming increasingly popular, personalization practices that involve building long-term *profiles* of the user (such as those used by Facebook and Google) are currently most common. Incorporating what is known as "collaborative filtering," these decision-making systems produce "user-specific recommendations based on historical user data" (Said & Bellogín, 2018, 98), which is then compared and contrasted to other user data to produce recommendations. Cohn (2019) highlights that recommender systems are an integral part of the contemporary web's sociocultural economy in ways that position "personal relevance" as a key signifier of cultural and monetary value, as I explore in the following. It is user-centered (rather than content-centered) personalization practices that take the interest of this book, precisely because they rely on personal user information to generate recommendations to users.

In addition to user-centered recommender systems, the other kind of algorithmic personalization most commonly intervening in users' daily experience is micro-targeted advertising, enacted through a practice known as Real-Time Bidding. Used to display personalized "banner" and "side" ads[3] on web pages across the internet, real-time bidding works as an auction process, wherein advertisers bid for an "impression" (ad space) seen by a particular user on the website she or he is visiting. Bidding, as the name suggests, is in real time and is largely fought and won using a combination of user profiling and content review of the website hosting the advertisement. As noted in the following, the user profiles compiled as part of data monetization processes can be detailed and highly granular, and work alongside other profiling mechanisms provided by first- and third-party data brokers in a complex process of user surveillance, management, and monetization. Though they certainly have the capacity to be complex, when it comes to advertising, profile-based auction systems often employ one or two key demographic markers in their attempt to reach the "target audience." So if, for example, a user who has been profiled as "female," aged "25–34" (among other more complex categories of interest and habit) chooses to watch a video on YouTube, an advertiser, for example promoting ClearBlue fertility and pregnancy tests, can bid in real time for their ad to be displayed to that user—deemed to be interested in the product because of their algorithmically inferred gender and age. The sociocultural implications of algorithmically assuming a user might be interested in fertility products because of their algorithmically inferred "gender" are performative and material, and open up questions surrounding what can be considered as "sensitive" information, as I explore in Chapter 3. However, for now it is important to note that real-time bidding is not unique to

YouTube: almost all online search advertising is sold through this process, because it allows marketers, brands, and businesses to deliver personalized ads based on user demographics to the same user across sites, devices, and platforms. Hailed as "the most significant progress in recent years in online display and advertising" (Wang et al., 2017, 1), real-time bidding's most significant attribute in regard to algorithmic personalization is that it heavily relies on "behavioural targeting" and "retargeting" (Deighton and Johnson, 2013, 44) of web users in order to deliver personalized advertisements.

With these common practices in mind, there are a couple of points I would like to make to set up some foundational characteristics of algorithmic personalization—characteristics that I argue are key to understanding the drive to personalize as a commercial "force relation" (Bucher, 2016) that acts upon web users at the level of everyday interaction. In most contemporary personalization practices, *profiling* is key to algorithmic personalization: the user's behaviors, click-throughs, inferred identity markers, and other signals must be profiled and categorized in order to establish what is of "personal relevance" to that user. As Christl and Spiekermann (2016) note, profiling mechanisms use a dizzyingly extensive list of categories in which users are placed: including gender, age, ethnicity, lifestyle and consumption preferences, language, hobbies, personality traits, location, political leanings, music and film taste, income, credit status, employment status, home ownership status, marital status—the list goes on. Despite the increasing detail of such profiling, and in the wake of the EU General Data Protect Regulation (GDPR), which stipulates that even an IP address or cookie identifiers can be considered as identity-marking technologies, platforms continue to perform intricate discursive wordplay to assure users of some level of anonymity in how their data are processed and brokered. For example, Google's Privacy Policy states, "We don't share information that personally identifies you with advertisers" (Google, 2018) suggesting a degree of anonymity that might sound convincing but that makes increasingly little sense considering the effect that profiling has on users' personal experience of the web, as I explain in the following.

Whether this profile is made up of long-term or short-term interaction with a site or set of sites, algorithmic personalization is premised on the idea that your *future preferences* can be inferred from your *past interactions*. Your previous click-throughs, likes, and browsing histories are assumed to be the best means of computationally inferring your present behaviors, and so your past web trajectory goes on to determine what you will see next. As established in Chapter 1, Pariser (2011) criticizes this key component of personalization, arguing that the assumption that past habits equate to future preference creates a hypothetical "you-loop" of digital experience for web users. He argues that personalization "can lead you down a road to a kind of informational determinism in which what you've clicked on in the past determines what you see next—a Web history you're doomed to repeat" (2011, 16). This, Pariser claims, has a negative impact on a user's sense of self: "[Y]ou can get stuck in a static, ever narrowing version of yourself—an endless you-loop" (2011, 16). As such, this key characteristic of personalization comes with a disconcerting implication of repetitive, reductive experience for the users who encounter it.

Pariser's (2011) critique certainly helps to highlight a detrimental assumption implicit in the idea that past action can be used to predict future behavior. However, the "you-loop" somewhat underplays another key component of personalization: that the "profile" built on you is *not* simply a static aggregation of your past habits and preferences. Instead, the profiles built by data trackers can be considered as complex constellation of data points, compiled through establishing "like-to-like users" and *only* rendered useful when aggregated with and against correlations to other groups of users. In monetization terms, this is far more sensible than relying on a single user's past interactions to infer her, his, or their future preferences. After all, the end goal of the "you-loop" if only based on a user's history would not be very profitable—the you-loop would become so repetitive as to omit the introduction of new and productive means of marketization. Instead, it is the categorization of the user *with and alongside other users* that apparently makes algorithmic personalization such an attractive market strategy. It is a point emphasized by Jarrett (2014), Gillespie (2014), Chun (2015), Cheney-Lippold (2017), and Cohn

(2019): your profile is only made meaningful and commodifiable to marketers in and alongside the context of other users' profiles.

This leads me to another crucial characteristic of algorithmic personalization. As already established, unlike other forms of personalization, it is algorithms that are responsible for determining what is of relevance to the individual, and so it is the "system" and not the "user" that is positioned as gatekeeper in the decision-making process. The decision-making capacity of algorithmic personalization systems is championed by advocates as time-saving, pleasurable, and convenient—for instance, Mayer-Schönberger and Ramge (2018) celebrate the ease of using adaptive personalization systems for deciding what information to prioritize, when to consume it, and how, as I will explore shortly. Of course, even when it is enacted in pleasurable or time-saving ways, consuming algorithmically personalized content or services still means allowing the algorithm to act in your stead. As such, algorithmic personalization inherently involves affording some autonomous control to the *algorithm as gatekeeper*.

It is important to highlight, however, that even when the system is given gatekeeping agency, the user still functions as a *data provider* for the system (Van Dijck, 2013): in order for the system to initiate any kind of personalization, the user must still provide some kind of data or input for the system to act as a decision-making gatekeeper. It should therefore not be assumed that even when the system is given priority over the personalization process that the user is not entangled with(in) the system—the user's data always inform the personalization process to which the user is subject, even when the user is not primarily in control of such processes.

Spanning all of these characteristics is the idea that *personal relevance* is key to enhancing a web user's experience. Of course, there are quantifiable benefits to having services algorithmically rendered "more relevant" to the user as an individual social subject: cookies streamline site visits by "remembering" user details, autofilling technologies allow users to quickly complete registration forms, and recommender systems can help navigate the otherwise unmanageable amount of content and services currently available via networked technologies. However, the idea that "personal relevance" unproblematically equates to better user experience is critically questionable. To begin with, as Gillespie argues, the term "'relevant' is a fluid and loaded judgment, as open to interpretation as some of the equivalent terms media scholars have already unpacked, like 'newsworthy' and 'popular'" (2014, 174). Van Couvering takes this further, observing that over the past few decades since the conception of search engines, "relevance has changed from some type of topical relevance based on an applied classification to something more subjective" (2007, 186). Differently put, contemporary notions of what is "relevant," at least in regard to digital practices, are now measured in relation to individual users, rather than to an idea of "objective" relevance (Scannell, 2005; Kant, 2014; Pariser, 2011).

In HCI research, achieving optimal personal relevance presents a somewhat insurmountable issue: there is an inherent "noise" of everyday life that precludes the possibility of achieving "perfect relevance" for any given individual user. Known as the "magic barrier" (Said & Bellogín, 2018), this point of optimal yet unattainable relevance represents the impossibilities of completely and perpetually satisfying a user's needs and desires. "Perfect satisfaction" can be very much considered an impossibility because an individual's needs, desires, and preferences are not static, universal, or definitive. Instead, they emerge from and iteratively change in relation to other individuals, as well as an individual's environment, her, his, or their situated context, and indeed more broadly, her, his or their positionality in time and space. It would seem, therefore, that there is an inherent limit to achieving a perfectly personalized system. For HCI researchers such as Said and Bellogín (2018), the magic barrier is certainly a foundational problem in the development of personalization technologies: but it is one that can be potentially overcome. They suggest that one possible solution is to model personalized recommender systems on sample users who show the most consistency in their preferences over time and across contexts, thus reducing the potential inaccuracies caused by users who display indecisiveness or changeability in their desires.

Though perhaps efficient, the idea of stabilizing relevance in this way is built on a homogenizing principle that looks to standardize relevance and, in doing so, negate the complexity and diversity of user intention. This form of homogeneity is increasingly being imposed not just through media recommender systems, but in the form of everyday "digital assistants" such as Google's, which I examine in Chapter 6. I do not mean to suggest that systems that *could* break the magic barrier are any less problematic: even theoretically, such systems present profound problems for the exercise of individual autonomy, as I will revisit shortly. What I do want to stress is that in HCI research that looks to achieve optimal personal relevance, there exists an assumption that personalization inarguably and unproblematically improves users' web experience. Personalization is framed as an instrumentally beneficial rather than critically questionable practice by those interested in its technological development and implementation.

There are two points I want to make here in regard to the characteristics of algorithmic personalization. The first is that algorithmic personalization, however it is implemented and by whichever algorithmic actor, is *always* designed to individually intervene in a users' web experience. For instance, the collected data that go into inferring the gender of a user eventually result in the user being delivered a "relevant ad" that is specific to the user's situated and located experience of the web. Even if the user data collected to personalize an experience are completely anonymous, even if they are only collected on "short-term" or "session-based" habits, the drive to personalize always has the end goal of intervening in and changing a person's individual experience of the web. This might seem like an obvious point, but it is an important one that is in danger of getting lost in platforms' claims of data anonymity, and even in critiques that emphasize (quite correctly) that digital marketers do not really care who you are (Einstein, 2017, 183). The individual interventions of algorithmic personalization emerge most poignantly in the participant accounts featured in this book, all of which suggest that personalization—anonymized or not, privacy-invading or not—changes these web users' engagement with their online trajectories—and indeed their engagement with their own sense of self.

My final point is not so much a "characteristic" of algorithmic personalization, but more a condition created by its implementation: and despite its inherent indefinability, it is pivotal in foregrounding the relevance of algorithmic personalization in everyday life. Given the ubiquity and complexity of the practice, it is almost impossible to know exactly what, when, and how web users encounter personalization. In fact, because of the "overwhelming" (Christl and Spiekermann, 2016, 7) presence of data tracking on the contemporary web, Christl and Spiekermann describe the intervention of personalization algorithms into the everyday as "Kafkaesque." It is worth quoting them at length here:

> People can at times confront a Kafkaesque experience. We don't know why we see a specific ad, why we receive a specific offer, or why we had to wait hours on the phone hotline. Was it because we acted in a specific way before? Did the newsletter of that political candidate contain issues, which were personalized in a specific way? Was it because we visited a specific website, used a specific mobile app, bought a specific product in the supermarket or watched a specific TV program? Could it happen that we get a loan denied someday, when we visit the online gambling website once too often today? Under the conditions of today's opaque and non-transparent networks of digital tracking **individuals do not know**, which data about their lives is recorded, analyzed and transferred–and which decisions are being made based on this information. (2016, 129)

It is this unnerving lack of clarity—the incompleteness of the knowledge that users can access and grasp about algorithmic personalization—that I argue emerges as *epistemic uncertainty* for web users who encounter personalized experiences. As Brunton and Nissenbaum assert, the complexity of data tracking creates a plethora of "known unknowns" (2015, 48) for users—users know that their data are collected, but do not know how their data are managed and processed. These in turn create "unknown unknowns" (Brunton and Nissenbaum, 2015, 48)—the ways in which data are processed and managed that users have no awareness of. These forms of epistemic uncertainty can affect future engagements, as Skeggs (2017) and O'Neil (2016) take pains to detail: users' data are used to predict their future desires and aspirations, to value them as consumers, or to devalue them as debt risks. However, Cheney-Lippold (2017) argues that data tracking also affects knowledge of the present—he argues that the ways we are made in data have performative implications for users' identities. I will explore this further in Chapter 3, but here I want to stress that algorithmic personalization creates epistemic uncertainty that can emerge as trust, anxiety, or somewhere in between, for users entangled within personalization systems.

A HISTORY OF THE ANTICIPATED USER

How did the algorithmic anticipation of users come to dominate the web as a market practice? To answer this question, it is useful to map the development of some of the technologies that have increasingly sought to "know" the individual.

As scholars such as Abbate (1999, 2010) and Campbell-Kelly and Garcia-Schwartz (2013) note, though it is commonly accepted that the internet has its roots in the US Defense Advanced Research Projects Agency (DARPA) military initiative of 1969, its subsequent development was driven by "a multitude of actors" (Abbate, 2010, 11) interested in the internet's implementation outside of military use. For example, US state-funded bodies such as the National Science Foundation (NSF) were interested in its value as a scientific tool, while commercial enterprises such as IBM saw its potential to facilitate corporate communication. Though these bodies are brought together through a number of narratives and counter-narratives currently in circulation, most historical accounts agree that from the late 1980s onward, the internet was subject to rapid moves away from publicly funded avenues of development toward development through privatization (Campbell-Kelly and Garcia-Swartz, 2013). Abbate argues that this privatization may have been driven by "technical necessity" (2010, 14) rather than commercial interest—however, it still stands that the originally state-funded internet was increasingly privatized by the Internet Service Providers (ISPs) who sought to roll out the internet to the general public through commercial enterprises.

At the same time this privatization was taking place, the world wide web[4] was being made freely available and becoming increasingly popular. Developed in the late 1980s by British computer scientist Tim Berners-Lee, the world wide web was from its inception accompanied by discourses that were largely oppositional to the privatization of the internet. These discourses maintained that the internet should be a *public*, *open*, and *free* resource, and again emerged historically through different actors—US politicians such as Al Gore wanted the net to be an open, widely available pubic service (Abbate, 2010); the scientific networks that the net originally fostered similarly wanted a collaborative, non-commercial network (Curran and Seaton, 2010); and finally, internet users themselves were also calling for the net to remain a public service (Abbate, 2010). Furthermore, according to Curran and Seaton, web founder Berners-Lee was also "inspired by two public service precepts: the need to create *free* public access to shared cultural resources . . . and the need to bring people into communion with each other" (2010, 263). Abbate (2010) observes that the rollout of the web was neatly timed to coincide with the launch of the commercial ISPs mentioned earlier—thus, though the internet itself was being privatized in some senses, a free and public-oriented web emerged in conjunction with commercially driven net developments.

Such narratives thus at least partially emphasize the "publicness," "freedom," and "commonality" that the internet and world wide web could potentially facilitate. As the uptake of networked home computers grew, celebrants of newly formed "virtual communities" such as Rheingold (1996) increasingly championed the collective and democratizing capacities of the web to foster collaborative, egalitarian spaces for public communication, free from commercial imperatives and encroachment. However, not everyone saw the increasing commercialization of the internet as antithetical to the web's emancipatory and egalitarian power. For example, business scholars Hagel and Armstrong (1997) disagreed with Rheingold and others that virtual communities should be inherently "anti-commercial" and instead argued that "profit motives will in fact enhance" both community interaction and individual consumer power (1997, x). Similarly, building on the ethics of libertarian technologists such as Brand

(1988), web celebrants such as Negroponte (1996) and Gates (1995) advocated that the commercialization of the web could bring about greater user power and increased individual freedom in the form of consumer choice and a more personalized interaction with content providers, businesses, and services. Negroponte (1996) envisioned that the web would function not only for the needs of the virtual community, but also the needs of individual web users by facilitating the creation of "boutique industries" that could cater to personal needs and desires. He further argued that the web could bring about a new form of personalized media consumption in the form of "The Daily Me" (1996)—an ideal of an acutely personalized newspaper that was not technically possible at the time but would one day deliver personally relevant news and entertainment to individuals in ways not possible through mass media and broadcast television. Other celebrants of the net's individualizing capabilities, such as Bill Gates (1995), saw similar potential in online services to provide computational "personal assistants" that would in the future automatically manage, suggest, and deliver "relevant" content based on individual preference.[5] In these arguments, the commercial benefits of an individualized web thus began to emerge alongside counter-discourses that celebrated networked technologies as collaborative and public.

As explored in Chapter 1, echoes of Negroponte's "The Daily Me," and Gate's "personal assistants" have all come to be (problematically) realized in some form. Cohn's (2019) analysis of the development of collaborative filtering technologies by MIT's Pattie Maes highlights that some of the first web personalization technologies to be commercially successful—such as film and music recommender Firefly, launched in 1995—were underpinned by a neoliberal ethos of sovereignty of choice that those such as Gates and Negroponte so passionately lauded. However, though personalized media recommenders such as Firefly were being developed as commercial successes, web-based *advertising* was still struggling to find its footing (Hagel and Armstrong, 1997; Turow, 2012; Curran and Seaton, 2010). Writing at the time, Hagel and Armstrong note that "commercial enterprises are relative newcomers to the online world, and so far few of them have made money" (1997, 3). They lament that

> [m]ost business on the Internet and other networks today do little more than advertise their wares on "billboards" on the World Wide Web . . . these old-media advertisers, dressed in new-media clothes, are only one indication that marketers have yet to discover the secret to unlocking the revolutionary potential of the internet. (1997, 3)

In other words, despite the "new media" marketing possibilities that the web seemed to offer, online platforms largely continued to turn to established forms of commerce in their attempts to profit from their web sites. For example, Turow (2012) argues that late 90s online banner ads functioned largely the same way that traditional ads worked, yet as Deighton and Johnson note, "so-called banner advertising of this era did not perform well by comparison to print or television advertising" (2013, 44)—from 2000 to 2003 online revenue through advertising actively fell.

Despite this reliance on "old media" marketing strategies, it did not go unnoticed that these banner ads did have one apparent advantage over established print media: they included the benefit that advertisers could tell when a user had apparently shown interest by clicking on their ad (now known as the "click-through" model). It is via the monitoring and management of click-throughs that data tracking began to emerge as a market practice. Crucially, however, as Peacock notes, at this time "data exchanges between Internet users' computers and a remote server were anonymous" (2014, 5). At this point, online advertisers could not personally identify or profile individuals—instead, "online ad space was based on proximity to content" (Cohen, 2013, 78) and it was the *audience or demographics* of a site that was sold as a commodity, rather than *individual user attention*.

As such, targeted web advertising in the mid-1990s web can be largely considered as synonymous with long-established strategies for "niche marketing" developed in TV broadcasting and print practices, which looked to segment the audience into particular group identities (Kant, 2014; Smith-Shomade, 2004). Audience targeting and categorizing are far from new: for example, Dixon and Gellman point out that individual profiling in the form of "identity scoring" dates back to 1941 (2014, 80), when the categorization of individuals started to be used to calculate the risk of lending credit to certain groups of people. Similarly, the practice of targeting individuals through niche audience segmentation and "static" customer identifiers such as addresses or phone numbers (Deighton and Johnson, 2013) is not something especially new. As early as the 1930s, marketers have sought to sell products not based on the function or form of commodities, but through a brand image or lifestyle ideal designed to appeal to certain classed, aged, and gendered demographics.

In the mid-1990s, then, if online service providers such as AOL, Prodigy, or CompuServe wanted to "know" their users, they had to adhere to established, "static" forms of identification historically common in marketing—for example, encouraging users to input personal data about themselves through registration forms and email addresses, or through explicitly registered transaction data (Hagel and Armstrong, 1997). That is not to suggest that these explicit forms of identity registration, profiling, and articulation did not come with critical considerations: as the work of Bassett (1997) and Killoran (2002) highlights, the restrictive interface and computational architectures provided by early platform providers sought to redirect users' actions and trajectories through systems of governance that can be critiqued as standardizing and reductive, as I expand on in Chapter 3. However, here I want to stress that though still subject to disciplinary form of registration, web users could only be identified through static, self-registered and collective forms of data aggregation.

By the mid- to late 1990s, these "old-media" market models were changing. Hagel and Armstrong (1997) paint an optimistic picture of the potentials of the internet to transform media marketing via data aggregation and extensive profiling as users moved from site to site. They argue that these new online, cross-platform market models would bring "enhanced ability to target" (1997, 11), predicting that "virtual communities will accumulate detailed profiles of members and their transaction histories, not only with a single vendor but with multiple vendors across an entire product category" (1997, 11). They go on to insist that online vendors can "aggressively" use individual data "to tailor products and to create product and service bundles" in order to "both expand the potential customer base and generate more revenue from each customer" (1997, 11).

Cohn (2019) argues that it was Mae's developments in collaborative filtering that can be seen as one of the first implementations of these ideals as put into practice. Cohn notes that Maes et al. developed a user profiling system—commercially launched as the Open Profiling Standard (OPS)—that could securely store and manage individuals' personal information and credit card details, allowing user profiles to be exchanged between vendors. With the development of OPS then emerged the possibility for the data tracking, anticipation strategies, and personalization of content and services that are being mobilized on the contemporary web.

However, there is one point at which both Hagel and Armstrong's rhetorical vision and OPS as a practical implementation differ from today's datafied landscape: user control over data. Hagel and Armstrong speculate that ownership of user profiles will ultimately lie in the hands of users themselves and "will be accessed by vendors *only on terms established by individual members*" (1997, 11, my emphasis). Cohn notes that the same emphasis on user privacy and control over data can be traced in the development of OPS, which was in part developed in response to fears concerning "the selling of this user information without meaningful consent" (2019, 76). He argues, however, that OPS was designed to *both* monetize user data while protecting users' right to control that data under the guise of privacy. Despite the monetizing of privacy at work, it remains interesting that, historically, user privacy was embedded in the developmental ethos of these technologies. Yet, as the works of Andrejevic (2013) and Jordan (2015) highlight, the drive for profit means that privacy measures have for at least a decade been very much set by *platforms* and not by users. Despite the GDPR allowing *some* user control over user data, data ownership on the contemporary web still largely lies in the hands of platform and service providers. I stress some because as I argue in the concluding chapter, the GDPR works to normalize commercial data tracking as an accepted part of web engagement, naturalizing the epistemic uncertainty inherent in data tracking even as it promises to give users more control.

As OPS developed, so too did ad networks such as AdSense, which allowed for the automated matching of website content to advertising content, meaning that even small "community" publishers and bloggers could begin to sell ad space without having to liaise directly with advertisers. Then, around the mid-2000s, came the development of the aforementioned real-time bidding. Allowing advertising to bid in auctions in real time for ad space meant that matching processes could be further streamlined; known as "Ad Exchange," real-time bidding facilitators such as RightMedia and the still hugely profitable DoubleClick (now owned by Google), began to emerge (Wang et al., 2017).

However as Deighton and Johnson note, "until this point [mid-2000s], while individual-level audience data were used to judge advertising performance *after* it has been bought, the data played no part in the buying process" (2013, 45). This changed in around 2007 when behavioral targeting began to be integrated into real-time bidding processes. In terms of algorithmic personalization, then, this is a crucial shift away from media *content* being the key to a successful sell (as with broadcast models) and toward users' inferred preferences and behaviors becoming the most important factor.

Underpinning all of these developments is the invention of the HTTP cookie. First developed in 1994 and used commercially by 1998, the cookie is broadly defined by Peacock as a small piece of code that facilitates "a way of storing information on the user's computer about a transaction between a user and a server that can be retrieved at a later date by the server" (2014, 5). The cookie became the primary data aggregation tool for commercial content and platform providers interested in knowing and anticipating their users. As Turow (2012) recognizes, in the early years of data tracking the cookie enabled web sites to monitor site visits and page click-throughs enacted by individual users, meaning that cookies marked the first move toward the individualization of what was formerly audience or demographic data.

Historically, user control over cookie placement on computers has been very limited: though there are a number of tracker-blocking softwares available to users interested in cookie prevention,[6] it is only since 2011 in the European Union that web sites have been legally required to notify users of data collection, with the 2018 GDPR giving users further (limited) control over what cookies to accept on their web browsers. Peacock acknowledges that user consent and control *were* at first taken seriously by the cookie's developers—in an initial paper that discussed the privacy implications for the tracking of users through cookie technologies, "[a]n extensive section is devoted to privacy problems and discussions on the rights of users to remove or generally cap cookies" (2014, 5). She highlights that, in these suggestions, "full agency is attributed to online users" (2014, 5). Crucially however, she notes that the measures implemented to protect users' privacy and agency were quickly removed—users lost the right to "remove and cap cookies" (2014, 5) before the technology even became widespread. Like the privacy ethos that Cohn identifies as part of OPS development (2019) and like the user control embedded in Hagel and Armstrong's (1997) profiling plans, the rights of users over cookie placements could not withstand platforms' drive to monetize user data.

It is through the development of the cookie in the late 1990s, combined with a market model that favored (targeted) advertising over pay-to-use service models, that the contemporary web economy starts to emerge. As Curran and Seaton note, the development of a privatized web in conjunction with narratives of public access and freedom "had accustomed people to expect softwares and content to be free" (2010, 253), rather than pay subscriptions or registration fees to access services, platforms, or content. In doing so, the development of the web economy followed a market model that relied on advertising as a primary revenue-generating strategy—and crucially, in regard to this book, a strategy that increasingly sought to anticipate, track, and target individual users rather than audience demographics. By the mid-2000s, the world's most-used web sites began to feature some of the key players still in play today—companies such as MSN, Google and later Facebook. Such companies could—and still do—attract large numbers of users while still retaining the attractive free-to-use model implemented by earlier (but less commercially successful) sites such as Geocities and Netscape. Thus free-to-use web services, funded by (increasingly personalized) marketing strategies rather than pay-to-use mechanisms, can be considered one of the key economic models currently driving the contemporary web.

As this web economy has developed, data tracking has become more and more complex. Peacock (2014), Wang et al. (2017), and Nikiforakis (2013) note that HTTP cookies are only one means of data tracking—many more technologies have developed since. Users are now pursued by embedded objects like "supercookies," "zombiecookies," "ubercookies," or "evercookies"—and "these tags are no exaggeration" since these new forms of cookie are "almost impossible to circumvent" (Peacock, 2014, 6). Similarly, Nikiforakis et al. (2013) reveal that commercial data trackers have a range of "cookieless" methods for identifying and anticipating users. Common types of tracking include Flash and canvas "fingerprinting," which are seen as preferential to cookie tracking as fewer web users are aware of these technologies and, unlike cookies, they cannot be easily deleted, meaning they can profile users for weeks or months after first installation (Wang et al., 2017). As Wang et al. report, the exploitation of cookieless tracking is increasing (2017, 15), suggesting that despite GDPR regulations there are still countless ways in which users *might* be tracked without their explicit knowledge or consent.

The historical development of the web's market model is important to this book for two reasons. The first is that by contextualizing current user entanglements with algorithmic anticipation, we can see that the data-for-services exchange is not inevitable: it is a historically and techno-culturally specific model. This market model—wherein users "pay" with data rather than money—has meant that commercial platforms can continue to embrace discourses of "publicness" by providing cost-free services and yet still generate profit by commodifying user data and interactions. As I have argued, this data-for-services exchange is very much at the heart of today's web economy: in fact, according to Mayer-Schönberger and Ramge (2018), data brokerage is set to replace money as the key to value calculation in late-capitalist free markets. This speculation is somewhat questionable, as I will review in the concluding chapters of this book; however, their assertion does help to highlight that the drive to personalize comes underpinned with data-tracking strategies that have implications for both the web economy and the lived experiences of web users. Second, this market model is significant to my research because current data-tracking practices have implications for how users are "known" by platforms, as I explore in the next section.

FROM UNIQUE TO "DIVIDUAL": MAKING SENSE OF THE DATA-TRACKED SELF

The rapid development of algorithmic personalization technologies has brought with it increasingly individualized means of targeting users. What is omitted from the preceding historical analysis, however, is an interrogation of how algorithmic identifications of "a user" can be considered to equate to knowing "a person"—that is, how the data collected by platform providers about a user's web trajectory are translated into identity markers that are suggestive of a person with (commodifiable) tastes, preferences, and habits. Though data trackers seem to suggest this process is straightforward, it is useful to briefly consider how the data signals collected by trackers are made sense of and monetized.

The proliferation of data-tracking technologies described earlier has not just meant more *ways* to identify and categorize individuals: algorithmic identification seeks to epistemologically "know" the world through computational ontologies and registers that can be considered a distinct form of knowledge production. Finn argues this point when he examines the ways in which computational cultures seek to epistemologically organize the world, arguing that algorithms act as ultimately always unknowable "cultural machines" upon which humans increasingly rely to produce knowledge (2017). Referring specifically to algorithmic profiling, scholars such as Cheney-Lippold (2017) and Bolin and Andersson Schwarz (2015) note that developments in data tracking have resulted in a scenario in which individuals are no longer "known" and targeted by the established identity markers and demographic models typically used by "old" media marketing. For example, Bolin and Andersson Schwarz point out that in place of the traditional data variables "such as age, gender, ethnicity, education [and] media preferences" (2015, 1) once used by marketers to discern the demographic qualities of groups of individuals, new technologies "register consumer choice, geographical position, web movement and behavioural information" (2015, 1). These new computational registers are wholly abstract: contemporary data-mining techniques attempt to capture mass data sets, which look not for representational information but for "pattern recognition" (Bolin and Andersson Shwarz, 2015, 4). Thus, what was once a matter of *representing* audiences by definition becomes a process of abstracted "*correlation*" (2015, 5, my emphasis). To quote Bolin and Andersson Schwarz at length:

> The explanatory dimension of representational statistics (e.g. "this group of people behave like this due to their social composition and their habitus privileging certain kinds of action over others") becomes less important than the establishment of correlations between (probable) behavioural patterns. The socially explainable "who" behind this pattern is less important than the algorithmically predictable behavioural "how." (2015, 5)

In contemporary personalization practices, then, audience identification is no longer a matter of representing people using established referents of "who they are" as social subjects—"old," "man," and "woman" are replaced as identification markers by purely computational referents of correlational and networked positionality.

Yet, if this is the case, this begs the question: Why are users given access to profiles—such as Google's and Facebook's Ad preference profiles—that present users with a gender, age range, and inferred but very much non-computational cultural interests ("feminist," "shopping," "R&B")? Bolin and Andersson Schwarz explain this, arguing that contemporary data-tracking strategies are (by definition) so abstract that data must be *translated back* into the "traditional social parameters" (2015, 1) that marketers have long understood and have for many years been used to commodify audiences. It is, after all, in the "translating back" of abstract computational ontologies that the tracking, managing, and anticipation of the individual becomes profitable. Data tracking is a costly exercise that only becomes value-generating if advertisers can convert data ontologies into categories that can inform and generate capital. As Andrejevic (2013) points out, these translations are not actually very "accurate"—only 75% of user inferences are considered "authentic" reflections of user identity and preference. However, from a data science and marketing standpoint, they are considered "close enough" to be profitable, despite the marginalization and cultural homogeneity that such translations might reinforce, as I explore in Chapter 3.

Cheney-Lippold's (2017) work helps further illuminate this shift toward wholly computational identifying markers. Cheney-Lippold theorizes that users are now constituted through algorithmically defined "measurable types"—a constellation of categorizations that adhere to established markers such as "man," "terrorist," "celebrity," but that are, as Andersson and Bolin Schwarz (2015) similarly assert, produced wholly in and through data. These data markers help to constitute what Cheney-Lippold calls "algorithmic identities"—material and performative entities based on "extant, datafied objects that can determine the discursive parameters of who we can (and cannot) be" (2017, 48). He stresses that, much like "offline" identity, the algorithmic self is not fixed or static: because of the recursive, correlationary, and abstract qualities of algorithmic datafication, "our measurable-type identities shift according to changes in data" (2017, 29). Cheney-Lippold proposes that the shifting and recursive nature of algorithmic profiling means that culturally established markers such as "gender" can become essentially dis-anchored from the "gender" that platforms such as Google and Facebook infer us to be.

The de-anchoring of cultural identity markers through algorithmic identity constitution can be considered as both problematic and productive, as I deliberate in the next chapter. However, here I want to emphasize that in order to identify and anticipate individuals for the sake of personalization, the "person" being tracked is essentially *disassembled* into a selection of data points. As Cheney-Lippold (2017) considers, we are no longer "individuals" within the system, but abstract and algorithmically manageable constellations of data. Ruppert et al. (2013) note furthermore that not only are people "disassembled into a set of specific transactions or interactions" but that "it may or may not happen that they are *reassembled* into people" (2013, 36, my emphasis). Thus, to speak of Google and Facebook tracking "you," as privacy advocates often speak of, becomes largely redundant. Commercial platforms seek to disassemble individuals into markers that can be isolated, stripped away from each other, and correlated alongside other users within categorical fields.

Such models thus very much constitute web users not as individuals and yet not as masses either. In doing so, and as Lyon (2014) and Cheney-Lippold (2017) recognize, they correspond to Deleuze's earlier (1992) theorization that computational mechanisms work to turn users into "dividuals"; that is, "we no longer find ourselves dealing with the mass/individual pair. Individuals have become '*dividuals*', and masses, samples, data, markets or '*banks*'" (1992, 5). I wholeheartedly echo the sentiment that the figure of the "dividual" encapsulates the fragmented, correlationary qualities of the data constellations that seemingly "haunt" users as they experience a personalized web: there is no individualized ghost in the machine, only abstract, recursive data assemblages that are "reassembled" to correspond to figures outside of the algorithm. For me, however, questions persist: How do the *dividual* constellations constituted by algorithmic personalization intersect and interact with the *individual* users who scroll through the interfaces that are changed and personalized as a result of dividuation? How is dividuation, despite its dehumanizing qualities, felt and experienced by the social subjects it is very much intended to affect at an individual level? It is a question that warrants an understanding of both people "as data" and whose lives continue to be lived out of the algorithm, as I will explore in the latter chapters of this book.

USER VERSUS SYSTEM: THE STRUGGLE FOR AUTONOMY

Though the individual is essentially dividualized by the drive to personalize, there remains a persistent discursive insistence that data tracking is focused around "you" the user. Whether legitimized by data trackers or resisted by privacy advocates, the self is often framed as central to the benefits/concerns that data tracking apparently produces. For instance, the commercial privacy policies used to explain data tracking claim it is undertaken to provide "you"—not a collective of users, not particular demographics, not the public—with a better experience. In order to understand the seemingly contradictory convergence of technological disassemblage with this discursive insistence of the unique self, it is useful to examine neoliberal discourses of web individualism that I argue have been used for the past decade to legitimize the privacy-invading practice of personal data tracking.

Discourses of web individualism have their roots in the dawn of "web 2.0."[7] Scholars such as Van Dijck (2009), Houtman et al. (2013), Jenkins (2006), and Benkler (2006) note that this next stage of the web brought with it a fresh set of apparent opportunities for the individual championed by free-market web advocates such as Gates and Negroponte in the early 1990s. At the time of its implementation in the mid-2000s, web 2.0 was popularly and critically celebrated as affording users more chance to produce, "prosume," and participate (Jenkins, 2006; Benkler, 2006). These popular rhetorics tended to champion new possibilities for web users to exercise individual autonomy over their web engagements: Houtman et al. emphasize that "[a]gency, personal autonomy and (inter)active control over new media content are at the heart of the new media's 'participatory culture'" (2013, 54).

From this participatory connectivity emerges what Rainie and Wellman call the "networked individual" (2014). Their model describes a situation in which traditional ideas of community ("virtual" or otherwise) based on kinship and collective interest become "segmented" and "sparsely knit" (2014, 135) in favor of the network—in which the *individual*—"not the household, kinship group, or work group . . . is the primary unit of connectivity" (2014, 124). This emphasis on the individual is, according to these writers, to be largely embraced—under this model, people are freed from the ties of traditional, institutional bonds (such as the neighborhood), and through personal hard work can reap the rewards of their own networked success. These networked individuals enjoy the autonomy of establishing their own individualized web experience—under networked individualism, "each persona also creates her own internet experiences, tailored to her needs" (Rainie and Wellman, 2014, 14).

Somewhat paradoxically, the networked individual comes to function as the figurehead of popular discourses surrounding "participatory culture." For instance, *Time* famously named "The Person of the Year" in 2006 to be "You"—celebrating the creative opportunities of this new collaborative environment through a lens that foregrounds participation as an individualistic pursuit. This emphasis on personal freedom and creativity does not exist in a vacuum: as I will explore in the following chapter, this idea corresponds to discourses of late-capitalist, post-feminist (Cohn, 2019) neoliberalism that celebrate the self as inner, unique, and an "agent of [its] own success" (Ringrose and Walkerdine, 2008, 227).

Despite (or indeed because of) this abundance of participation, a particular cognitive problem emerges: how to process and keep pace with the increasing informational activities generated by both the user as prosumer and the billions of others now liking, creating, sharing. With increased participation comes an increase in the need for informational decision-making—of what to prosume, how, and when. Scholars such as Lovink (2011) and Andrejevic (2013) describe a new mode of mediated existence: data-driven information overload, an "infoglut" (Andrejevic, 2013) wrought by ubiquitous and endless information production and processing, wherein "staying on top" of information becomes an epistemological impossibility. Andrejevic notes that as individuals, information becomes somewhat impossible for users to manage, meaning that "at the very moment when we have the technology available to inform ourselves as never before, we are simultaneously and compellingly confronted with the impossibility of ever being *fully* informed" (2013, 2).

How then to deal with this infoglut? For proponents of AI, the technological solution to this epistemological problem comes in the form of algorithmic decision-making. These adaptive and autonomous systems are manifest in a plethora of forms: from news feeds that can decide what is relevant on behalf of the user, to information and entertainment recommender systems, to everyday personal assistants such as Echo and Google Home. As these assistants increasingly move into domestic spaces, they also shift from collective informational retrieval (such as Google Search) to personal and everyday lifestyle management.

Mayer-Schönberger and Ramge (2018) celebrate the potential of algorithmic personalization technologies to help manage the everyday effects of information overload. Citing the "adaptive systems" developed to deliver product recommendations, they argue that personalization algorithms will become "our trusted assistants" that will not only be able to "do the boring stuff and reserve those decisions that give us the most joy and pleasure for ourselves" (2018, 85) but also protect us from making bad decisions based on our "biases." For these authors the autonomous networked individual is simply aided in decision-making by algorithmic personalization, to the extent that "eventually we may be able to delegate to adaptive decision systems some of the decisions we fret over" (2018, 218–219). As an *almost* cautionary caveat, they speculate that such systems "may even end up making a lot of decisions for us" (2018, 81).

Some scholars are a little more tentative in their engagements with algorithmic decision-making. Campanelli notes that the "freeing" of decision-making to computational systems assumes that "information selection should not be considered a core activity in the lives of human beings, but a burden that one may well leave to machines and their algorithms" (2014, 43). Hillis et al. (2013) take this further, arguing that imagined, acutely personalized information retrieval systems imply that "achieving perfect relevance would be akin to the technology seeming to read one's mind" (2013, 55). These theorizations help to highlight that the convenience of preemptive information management in fact "promises a limited form of virtual sovereignty" (Hillis et al., 2013, 22). Cohn (2019) argues that in the apparent "crisis" that is information overload, algorithmically personalized recommendations "offer relative safety and stability, and, in so doing, present real choice as an unsafe, potentially costly burden" (2019, 40). Though what can be considered as meaningful or "real" choice is something of a point of contention for me, as I will explore in Chapter 7,

I similarly view decision-making agents with skepticism in thinking of them simply as "assistive."

If we consider, as these scholars do, that decision-making is fundamental to social subjectivity, the relationship between the autonomous capabilities of user and system becomes increasingly complex. As I stated at the beginning of this chapter, personalization algorithms in their very purpose are designed to act as gatekeeper: to act in the users' stead in deciding what information or services are "relevant" to that user, and in doing so take on a role as an autonomous actor. But what does autonomy mean here exactly? Smithers (1997) argues that though HCI research tends to treat the idea of autonomous non-human agents as uncomplicated, defining autonomy in social actors—both human and non-humans—is a difficult task. Broadly speaking, I take the term to mean the ability to self-govern in decision-making. It is the ability of personalization algorithms to at least appear to make decisions in our stead that distinguishes them from other non-human agents—as Latour's (2005) work exemplifies, even doors or light bulbs can be considered to have agency in that they have the power to act on their surroundings, but this does not necessarily mean they are self-governing in their decision-making. Of course, in the era of the "internet of things," even networked doors and light bulbs are becoming "smart" enough to be considered autonomous agents, but this is largely due to algorithmic decision-making capacities. Smithers makes a compelling point about this distinction, arguing that *appearance* is key, and stating that autonomy is an "attribute . . . given to [an actor] by another observer" (1997, 92). Personalization algorithms differ from other non-human actors, then, precisely because we attribute to them the ability to make decisions in our stead—to recommend our next playlist, to filter our News Feeds, to deem events important enough to be input into a user's mobile calendar.

The apparent autonomy of algorithmic personalization technologies is a notion I will return to in Chapter 6. Here I want to point out that as much as autonomy is an attribute defined by the observer, so too can it be considered ideological and culturally constructed. Cohn (2019) and Noble (2018) note, for instance, that the workings of contemporary recommender and search technologies reflect the long-standing conflation of middle-class, white, male autonomy with the sovereignty of human subject. I argue in Chapter 6 that the "ideal user" of personalization apps such as the Google mobile app is built on similar dominant hegemonic norms—as I will explore, the ideal user of such technologies is assumed to be interested in informational daily trajectories of jet-setting, male sports teams, working, commuting, and stock brokerage in ways that align the autonomous social subject, in need of algorithmic assistance in daily decision-making, as a normative subject that upholds dominant hierarchies of race, gender, and class.

However, as much as algorithmic personalization implicitly reinforces the sovereignty of this normative subjectivity, I argue that it also undermines the social subject as sovereign agent; that is, in the unburdening of "boring" or "difficult" decisions, personalization algorithms *undermine* the very autonomy that is supposedly only being assisted. After all, the idea of asking an algorithmic system to act autonomously on a user's behalf is in some ways a paradox—the system cannot act autonomously for the user-as-agent without becoming an autonomous agent itself. In allowing algorithms to decide in the users' stead how to manage and process the infoglut of the everyday, the autonomous freedoms of the (networked) sovereign self is called into question.

It is the autonomous capacities of algorithmic personalization that open up critical considerations in regard to the decision-making processes that are at times "given over" to the system in the name of convenience. I propose that the autonomous capacities of algorithms create not just convenient "personal relevance"—or indeed feelings of privacy invasion—but also *a struggle for autonomy between user and system.* This is a struggle that decides what the user sees, what services the user has access to, and how information about the world, as well as the user themselves, is presented online. As I will revisit in the concluding chapter, the parameters for the struggle for autonomy continue to be redrawn even as the drive to commodify user intention persists. Traversing these parameters are the individuals who must negotiate the promises and pitfalls of algorithmic personalization, as I detail in the second half of this book.

I will return to this struggle at various points throughout this book, but especially in Chapter 5, where the autonomous capacities of Facebook's autoposting apps actually work to *undermine and disrupt* participants' self-performance on Facebook, and in Chapter 6, where I explore the capacity of Google's mobile app to autonomously deliver to users "the information you need throughout your day, *before you even ask*" (Google, 2016). Such predictive promise encapsulates the blurring of sovereignty between user and algorithm produced by algorithmic personalization, and the struggle for autonomy that is lived and felt in different ways by different users.

I conclude this chapter, then, by stressing that the contemporary data-tracked individual is not just anticipated and acted on, but also acted *for* by algorithmic personalization in ways that create space for negotiation between user and system in regard to knowledge production, control, privacy, (the struggle for) autonomy, and identity articulation. I will explore the nuanced implications of the drive to personalize in the latter chapters of this book, where the testimonies of users themselves become key in understanding the inventions of algorithmic personalization in everyday life. First, however, I turn to defining the "person" dividuated, performatively constituted, and confronted by algorithmic personalization.

3

Me, Myself, and the Algorithm

As the previous chapter has detailed, the processes of algorithmic personalization involve "knowing" and anticipating the user for monetization purposes. This is achieved through technologies that seek to dividuate the individual and position the algorithm as an autonomous agent—even as platform providers celebrate the autonomy of "you," the user. However, though the political economy approach adopted in the previous chapter illuminates algorithmic personalization as a marketized "force relation" (Bucher, 2016, 39), it is less helpful for understanding the social subject entangled in personalization systems. After all, if "personalization" involves rendering something personally relevant to an individual, then how is that "person" understood, constituted, and theorized? To explore this question, I want to scrutinize not just "algorithmic" identity, but also broader concepts of what is meant by personhood and the self. This chapter looks to understand how users have come to be understood both inside and outside of algorithmic constitution, and in doing so, historically and critically contextualize the participant negotiations that underpin the following three chapters of this book. Once again I offer a

Making It Personal. Tanya Kant, Oxford University Press (2020). © Oxford University Press.
DOI: 10.1093/oso/9780190905088.001.0001

historical overview, not of data tracking, but of the "individual" as it has been come to understood in both "online" and "offline" contexts. I argue that it is only through understanding social subjects as constituted both inside and outside of algorithmic protocols that it is possible to untangle the nuanced sociopolitical implications that algorithmic personalization produces in cultural interaction and identity practice. I also use this as an opportunity to think in more detail about how researchers constitute and negotiate the lived experiences that they study.

Given the huge breadth of work surrounding terms such as the "self," "personhood," the "individual," and "identity" (all distinctive formations in their own right), the following analysis can only ever be partial, but does give a foregrounding of some of the theories useful in understanding how algorithms might intervene in—but not necessarily determine—the horizon of possibilities that structure the (networked) everyday.

IDENTITIES: INNER, AGENTIAL, PERFORMATIVE

As established, the current drive to personalize is built around the premise that individual—as supposed to collective—need, desire, or relevance is of most value in terms of delivering to users an "optimal" web experience. In many ways, this emphasis on the individual is nothing new: "the self" as a concept has functioned as a core focus within neoliberal capitalist societies that take the sovereignty of the individual to be foundational to labor relations, sociality, and systems of production/consumption. Scholars such as Stone (1995), Geertz ([1979] 1993), Foucault (1988), Boellstorff (2008), and Rose (1991) acknowledge that "the notion of the self as we know it" (Stone, 1995, 89) within such contexts is largely underpinned by a particular formation of the individual as *inner, unique*, and *unified*.

According to Rose, this unitary and internal self is historically considered to emerge from Christian doctrine. These doctrines, themselves built on Ancient Roman concepts "of judicial and political personality," construct a social subject with an "internal existence" that could be awarded or burdened with—as well as disciplined through—a moral and religious conscience (1991, 221). What is culturally considered to be moral conscience has changed over time, as have the means used to express the inner workings of the mind. Foucault's work on "the technologies of the self" (1988) especially stresses that the self is always shaped and legitimized in historically specific ways. However, the self as inner and capable of morality is a notion that has remained as largely consistent.

Stone stresses that this notion of an inner and moralistic self might be persistent but should not to be taken as universal—the modern Western self adheres to a "classic bourgeois worldview," which she proposes emerged around the late 1600s and formed part of a political and epistemological shift that saw the world structured "into the form of binary oppositions: body/mind, self/society, male/female and so on" (1995, 89). These binary oppositions, especially the Cartesian view that the mind existed metaphysically separate from the body, gave rise to the idea, developed through the seventeenth and eighteenth centuries, that the body housed a conscious, rational, and unitary identity (Stone, 1995). Geertz asserts that this "Western conception of the person" is thus

> a bounded, unique, more or less integrated motivational and cognitive universe, a dynamic centre of awareness, emotion, judgement and action organized into a distinctive whole. ([1979] 1993, 59)

The increasing importance of the self as a "dynamic center of awareness" is reflected in the appearance of scientific disciplines interested in the individual as the central object of study. For example, the development of psychology in the late 1800s and early 1900s (Mattelart and Mattelart, 1998; Boellstorff, 2008; Rose, 1991) looked to uncover the inner workings of the mind as central to understanding human behavior.

Even as these developing disciplines sought to "unlock" the internal self, they also began to acknowledge the role that *social interaction* had in the formation of the individual. For instance, the 1920s marked the development of ethnographic approaches to self-constitution by institutions such as the Chicago School (Mattelart and Mattelart, 1998). From the 1940s onward, theorists such as Weber (1947) increasingly foregrounded the importance of group interactions for constituting the identities of the individuals that formed that group. The work of early ethnologists such as Strauss recognized that

> group membership is . . . a symbolic, not a physical matter, and the symbols which arise during the life of the group are in turn, internalized by the members and affect their individual acts. (Strauss, cited in Stone, 1995, 87)

With the notion that individual identities were constructed through group interaction came the increasing questioning of the assumption that the self was a unified whole that preexisted such social interaction. Notably, Goffman (1959) proposed that we "perform" multiple selfhoods that are called on and enacted depending on a person's specific social situation, and in doing so further critiqued long-standing notions that the self was necessarily unitary, stable, and static.

Work on the self increasingly sought to interrogate the relationship between the agential capacities of social subjects and the operational structures within which these subjects found themselves embedded. Foucault sought to position "the body" as constituted and disciplined by and within specific historical socio-contexts, in which "the body is the inscribed surface of events" (1988a, 1480)—meaning that, for Foucault, the human body constitutes a site of culturally contested meanings upon which discourses of governance and power are imposed, fought, and written (1988a). Hacking's work on "Making Up People" (1986) also emphasized the importance that discourse plays in how individuals are seen and managed as social subjects. For Hacking, the identifying and placing of people into social and economic categories are pivotal to the kinds of possibilities and limitations thus afforded to those categorized. Referring to institutionalized modes of managing people, such as census statistics and other state-legislated means of identifying individuals within populations, Hacking notes that "counting is no mere report of developments.

It elaborately, often philanthropically, creates new ways for people to be" (1986, 222). Though writing about eighteenth-century state measurements of national populations, Hacking's sentiments continue to resonate in contemporary claims that categorization is not just descriptive but *constitutive*: as Cheney-Lippold (2017) argues, the categorizations used by Facebook and Google to "know" their users do not simply identify individuals through algorithmic protocols, they produce the identities they categorize.

During the 1980s, debates around the ways in which the individual might be governed, disciplined, and determined by sociocultural discourses and structures gave rise to an renewed interest in individual will and agency—agencies that, through new regimes of sociality and consumption, have more power to change their lived circumstances and act as they see fit. Throughout this decade, theorists such as Butler (1988), De Certeau (1984), Hall (1988), and Hacking (1986) increasingly deliberated the agential powers of the individual who is always produced by discursive and structural operations, but *not* always governed and determined by them. De Certeau's work especially illuminates the intricacies of how individuals might find ways to deploy their agential capacities under the apparently oppressive conditions imposed by late capitalism. For De Certeau (1984), navigating the conditions of daily life involves a negotiation between the strategists (organizations, institutions, and those in power capable of structuring the everyday) and tacticians (those individual social subjects who find ways to "make do" through politically productive and culturally meaningful actions). "Limited by the possibilities of the moment," writes De Certeau, "a tactic is determined by the *absence of power* just as a strategy is organized by the postulation of power" (1984, 39) in everyday actions, meaning that the tacticians employ these maneuvers to find spaces of creation, resistance, and play in structures defined by strategists. For De Certeau, even actions such as the consumption of mass-produced goods (condemned by other theorists as upholding dominant power structures) can become a "a cultural creation, a poesis . . . a making and a using, a signifying constellation" (Poster, 2006, 413). As I explore in later chapters, the tactical engagements of social subjects take on a particular importance in relation to the anticipatory strategies deployed by personalization algorithms.

Other theorists such as Lasch focused on the increasing importance of individualism in late-capitalist societies in regard to both excessive consumption and self-centric ideologies that produced what Lasch called "the culture of narcissism" (1979). The idea that the self constituted a highly individualized yet malleable performance thus led to further—rather different—developments in how the self was theorized; for example, Gidden's work treats selfhood in twenty-first-century modernity as a "reflexive project," in which the self can be continuously and reflexively reworked (1991). Beck (1992) similarly argues that though not free from the confines of the social and economic structures that define life, individuals are free within these structures to set their own complex biographical narratives in and under the opportunities made possible by "reflexive modernity."

However, with the celebration of flexible, self-fashioned autonomy comes a critical counter-claim: an overemphasis on individual agency reinforces a focus on the self that ultimately neglects, and indeed reinforces, the unequal distribution of power and privilege in everyday life. As Skeggs (2004) critiques, the agency celebrated by Beck and Giddens fails to acknowledge or confront the class, gender, and race dynamics that exert force on the modern, empowered self. She emphatically insists that such formations function as "projects for intellectual aggrandizement" that ultimately "help to reproduce class inequalities more intensely" (2004, 54).

Other feminist scholars have sought to question the neoliberal assertion that in late-capitalist societies, individuals should be celebrated and indeed held responsible as "agents of their own success" (Ringrose and Walkerdine, 2008, 228). I join these scholars in their efforts to look for alternative models of theorizing the self—models that look to problematize the individual as a source of empowerment without denying that as social subjects, web users can and do exercise their choice and agency. Skeggs looks to counter neoliberal theorizations with models that instead posit an idea of the self as "produced through encounters with others located within relations of production and reproduction" (2011, 508). Differently put, it is by positioning the self primarily as entangled within relations with other social subjects, as well as within relations of production and consumption, that neoliberal celebrations of individualism can be interrogated as ideological. The social structures of late capitalism mean that one's successes are very much contingent on socioeconomic hierarchies that often uphold dominant class and power relations. Again, this does not mean dismissing the value of agency as something wholly ideologically determined: as the following chapters reveal, there is plenty of scope, even in algorithmically entangled identity construction, for the self to co-constitute and navigate its social relations. It does mean, however, paying intersectional, feminist, and class-conscious attention to the *kinds* of selfhood taken to be the most agential in contemporary formations of the self, as well as recognizing that individual agency can be difficult to locate and isolate in increasingly automated social networks.

Mobilizing concepts of "capital," Bourdieu's ([1979] 1989; [1984] 1998) work maps the ways that symbolic and material identity markers of taste, social status, and education (for example, one's taste in music, choice of car, and university attended) work to distinguish and differentiate between different classes and groups of people. These classed, gendered, and raced distinctions (among other markers) are what, according to Bourdieu, legitimize and delegitimize social subjects as "respectable" or "disreputable persons" in ways that uphold the power of elite social groups—for example, one's taste in "good" opera can be used as cultural capital to reinforce one's apparently natural higher status, while another's "poor" taste in pop music is used to explain and legitimize a lower social positioning. I unpick these notions up in Chapters 5 and 6, where I argue that algorithms are now intervening in the production of cultural capital, meaning that web users are increasingly being compelled to deploy forms of *algorithmic capital* to legitimize their own online actions, while delegitimizing others.

It is through gender theorist Judith Butler's work (1988a, 1988, 1990, 1998) that the apparently innate qualities of the "inner self" have come to be most recognizably questioned. Butler sought to build on Goffman's notion that identity can be considered as "performance," but argues that such performances are not exterior to our sense of selfhood; rather, they produce and constitute that very self. Butler's work thus corresponds to Foucault in that the self is produced through discourse, but also marks a departure in that "the body" is not simply inscribed with it (for this, Butler argues, paradoxically suggests the body exists outside the discourse that produces it) but is instead intrinsically "performative." In her now well-established theory of how gender identity might come to be constituted, Butler states:

> [g]ender cannot be understood as a role which either expresses or disguises an interior "self,"... As performance which is performative, gender is an "act," broadly construed, which constructs the social fiction of its own psychological interiority. As opposed to a view such as Erving Goffman's which posits a self which assumes and exchanges various "roles" within the complex social expectations of the "game" of modern life, I am suggesting that this self is not only irretrievably "outside," constituted in social discourse, but that the ascription of interiority is itself a publically regulated and sanctioned form of essence fabrication. (1988, 528)

For Butler, then, acts of identity performance are not just exterior acts, but are materialized and therefore constitutive of the very identity that they seek to perform. Butler's theories thus throw the idea that identity is unified, stable, and somehow preexists cultural construction very much into question. This book therefore takes Butler's notion of performativity as a central concern, yet also recognizes (as Butler herself does) that traditional notions of selfhood as holistic, inner, and unified still strongly resonate in contemporary popular treatment of the self.

In these theorizations, it becomes apparent that to "do" identity is a complex, fluid, and material set of acts—acts that entangle social subjects in processes that value and devalue certain embodied situated subjectivities as part of everyday life. As the following sections highlight, the performative implications for selfhood become especially pivotal when algorithmic actors intervene in identity construction. However, before exploring how this performative algorithmic self has come to be understood, I want to spend some time contextualizing the self more broadly within computational frameworks.

THE EARLY NET AND ONLINE IDENTITY

The development of the internet and subsequent world wide web outlined in the previous chapter has brought with it new opportunities to debate, discuss, and theorize identity in a space that appears to separate body from self. In the 1990s, when the web was in its infancy, writers such as Turkle (1997), Stone (1995), Rheingold (1996), Roberts and Parks (1999), Kennedy (1999), O'Brien (1999), Donath (1999), and Bassett (1997) explored the potentials and problems of articulating, maintaining, and (re)constituting identity in the "meatless" environment of the text-based "virtual communities" and other online forums that the web at that time facilitated. At the time, Turkle approached forms of web-based communication as an opportunity "for discovering who one is and wishes to be" (1997, 184), celebrating the fact that "[w]hen we step through the screen into virtual communities, we reconstruct our identities on the other side of the looking glass" (1997, 177). Similarly, Rheingold also described these new "virtual communities" as "a place where identities are fluid . . . we reduce and encode our identities as words on a screen, decode and unpack the identities of others" (1996, 61).

It is tempting to retrospectively structure a holistic narrative that early web theory was dominated by celebrations of the free and fluid self. However, critical debates of that time complicate this—the "freedom" this new virtual space afforded was always treated as politically questionable. For example, in her work on online gender expression, Bassett notes that the articulation of identity "beyond the flesh" (but *not* necessarily "beyond the body") created a "tension between gender play on the one hand, and a fairly rigid adherence to gender norms on the other" (1997, 538). Such theorizations, along with those of Burkhalter (1999) and Nakamura

(2002), emphasize that though the self might be separated from the physical site of the flesh, the culturally coded body continued to be legitimized or delegitimized even in cyberspace because it "emerge[s] through the power structures and gender asymmetries operating in Real Life" (Bassett, 1997, 538). These writers raise questions about the sociocultural codes, pressures, and ideologies that restricted and reinforced identity configurations both "online" and "offline," highlighting that leaving your physical attributes behind when communicating has never been enough to guarantee an "equal" exchange of bodies. As Noble's work (2018) highlights, the assumption that the disembodied self can equalize human interactions continues to be reinforced by technological frameworks that assume an "ideal user," as I explore in the following.

As outlined in Chapter 2, during the early to mid-1990s, individualized data tracking was still in its embryonic stages—internet users could not be tracked or implicitly identified by platform providers, and their individual habits, preferences, and identities could not be inferred to the depth and complexity that they are in today's landscape. It is here, then, that I want to emphasize the historical specificity of this context: articulations of online identity in this era of the net's history were debated in a sociocultural situation in which individuals *could not be computationally tracked and algorithmically categorized* in the ways that they are now. That is, though platform providers actively sought to attract and in some ways commodify user interactions, platform and content providers simply could not algorithmically identify individual users. Instead, providers had to either categorize such users via traditional advertising demographics, or users had to (mis)identify themselves or (mis)identify other users. The commercial rise of data tracking and the subsequent drive to personalize the web that I mapped in the previous chapter can be seen as having a profound impact on the possibilities of identity enabled by the web. Platforms' contemporary attempts to anticipate, and intervene in identity performance increasingly seek to pin down and anchor the self, even as they demand endlessly abundant (Szulc, 2018), recursively commodifiable identity articulation, as I will explore shortly.

THE IDEAL USER

Given the technological limitations of the time, users could not be identified as they are today. However, I do not mean to suggest that the user did not "matter" to platform providers: the figure of the (albeit undefined) user was taken as important to understand and cater to, even though the embodied social subject might not have been "knowable" through algorithmic means. As Oushdoorn et al. note, the late 1980s and early 1990s marked a shift in paradigms of design theory away from "technology-oriented design" and toward "user-oriented design" (2004, 30), wherein the needs of the users, rather than the producer's intentions, were placed at the heart of the design process. Thus in cyberspace the user might be unknown but could still be addressed: not as a specific individual, but instead in the notion of "ideal user"—an individual assumed to be "be everybody" (Oudshoorn et al., 2004), that could represent universal needs and preferences.

As Noble's (2018) and Cohn's (2019) analyses reflect, the position of the "ideal user" as a universal human subject has its roots in neoliberal free market ideologies. These ideologies have classically upheld the notion that if an individual can somehow be stripped of identity markers such as social status, class, age, gender, and wealth, then that person can participate as a "neutral" and unbiased candidate in democratic decision-making. This unmarked body—ungendered, unclassed, stripped of rank—is lauded as "the final goal of human transcendence" (Noble, 2018, 62) and can thus be trusted to contribute to a public good not based on one's own identity-specific interests, but an egalitarian ideal unsullied by situated subjectivity.

This universal subjectivity might at first seem to find its ideal home in the "meatless" cyberspace that individual web users could apparently inhabit anonymously. However, much like the production of "universal" knowledge explored in Chapter 1, the universal subject constructed as ideal has been critiqued as an ideological figure that, contrary to claims of neutrality, actually upholds dominant hegemonic frameworks of power and inequality. Noble describes the ideological subtext that accompanies the (supposedly) technologically realized universal human:

> This subtext is an important part of the narrative that somehow personal liberties can be realized through technology because of its ability to supposedly strip us of our specifics and make us equal. We know, of course, that nothing could be further from the truth. Just ask the women of #Gamergate and observe the ways that racist, sexist, and homophobic comments and trolling occur every minute of every hour of every day on the web. (2018, 62–63)

Essentially, then, the "ideal user" is not the neutral and unmarked figure it is celebrated to be—this imagined user is white, male, middle-class, and heterosexual. Nakamura (2002) and Oudshoorn et al. (2004) echo these sentiments; their work very much questions the idea that fulfilling the needs of the unmarked "ideal user" is possible, or even favorable, noting instead that normative identity markers function as an "invisible" default that reinforces dominant sociocultural hegemonies of who the user is and what the user should be. Noble (2018) argues that Google continues to foreground this ideal subject—along with its accompanying structural inequalities—as the assumed ideal user of Google Search. Cohn's (2019) work on recommender systems helps to unpack that the idea of freedom of choice in itself is ideologically tied to the white, middle-class male as the default and ideal form of agency, as I outlined in Chapter 2. My own work finds that the specter of "ideal user" continues to inform and determine Google's apparently "personalized" mobile app technologies, as I argue in Chapter 6.

The increased reach and ubiquity of data tracking, designed to know the "authentic" user at the end of the device, is now a commonplace and somewhat unavoidable characteristic of the contemporary web. The techno-commercial implementation of data tracking begs the question: Is the problematic figure of the "ideal user" now redundant? The increase of algorithmic personalization practices might seem to facilitate a move *away* from crude frameworks of use that construct the "ideal user" as a kind of universal, invisible, neutral subject. According to the platforms that sell personalization to their users, increasingly ubiquitous data tracking means the user no longer needs be imagined or speculated. Instead, service providers can now claim to have "real" access to *your* ideal use of a platform—for example, by tracking your "likes" and click-throughs, Facebook claims to individually tailor its service not for the assumed user, but for you as a person. Identity markers of gender, age, nationality, and race seem to have become not encumbrances of human transcendence, but the very key to providing the best experiences the online social world can give. However, as I will explore shortly, the idea that the "ideal user" has been replaced with an authentic and specific "you" can and is being increasingly questioned—especially if ideas of the performative online self are taken into account.

PROFILES AND PERFORMATIVITY

As scholars such as Sauter (2013) note, a plethora of potential "technologies of the self"[1] (Foucault, 1988) are embedded within the operational structures of numerous digital and social media platforms, designed to offer users opportunities to "express themselves." In today's contemporary landscape, status updates, "stories," live streams, videos, uploading photos, sharing links, location tagging, and "likes" (among other practices) are all commonly used to articulate identity markers and to construct the self. Advocates such as Miller (2011) celebrate such technologies as widening the possibilities of individual expression, noting that through social media profiles and other forms of self-generated content, social subjects are empowered in telling their own stories and presenting themselves in ways not possible through broadcast media.

Not all researchers share this enthusiasm—scholars such as Szulc (2018), Marwick (2013), Cover (2012), Thumin (2015), and Killoran (2002) note that contemporary technologies of the self can be considered to be empowering, but also as standardizing and restrictive. Thumin (2015) asserts that though social media has certainly democratized opportunities to tell your own stories, self-presentation has come to function as a condition for online participation: web users *must* create an identity profile in order to access many social media sites. Marwick further points out that social media profile formats (such as Facebook's) adhere to a fairly rigid framework that allows users to post status updates or photos, or log their favorite movies, but does not let users change the format, layout, or aesthetic of their profile (2013, 14). Thus, though self-expression is indeed permitted, it is restricted and conditioned by the structural architecture of the platform.

As social media networks (SNSs) have developed over the years, there has been arguably more and more choice (if not flexibility) built into platform architectures: notable examples include Facebook's adding of more "reactions" to accompany the "like" button and Twitter extending its Tweet character count to allow users apparently more freedom in the kinds of articulations they can create within the confines of SNS formats. Gerlitz and Helmond (2013) speculate that Facebook's third-party apps—explored in Chapter 5 of this book—might further facilitate an expressive framework that extends beyond merely being able to "like," "laugh" at, or "recommend" something; instead, users can increasingly "read," "watch," "discuss," or "perform other actions" (2013, 1353).

It is worth stressing that these emerging identity affordances are formatted in ways that adhere to data-driven algorithmic logics. Though they seem to offer increased possibilities for self-expression, these possibilities exist within a framework that easily facilitates the behavioral profiling which, as I established in Chapter 2, currently underpins commercial personalization practices. For instance, though they undoubtedly allow users the chance to more than "like" something, Facebook's new "reaction" buttons can still be easily tracked, mined, and managed in ways that can be commodified and used for the micro-targeting on which Facebook relies to generate revenue.

To elaborative, affective identity articulations such as those facilitated by Facebook's "reactions" buttons can be considered as expressible increasingly through strictly machine-readable frameworks. Agre (1994) calls these modes of structuration "grammars of action" that, through a "capture model" (rather than "surveillance model," as per Foucauldian models) of socio-technical governance can determine, regulate, and control the horizon of possible actions that web users can take. He writes:

> [o]nce a grammar of action has been imposed upon an activity, the discrete units and individual episodes of the activity are more readily identified, verified, counted, measured, compared, represented, rearranged, contracted for, and evaluated in terms of economic efficiency. (1994, 119)

According to Agre, in being more "readily identified" by algorithms, these activities are thus put through a process of "instrumentation that entails the reorganization of existing activities" (1994, 122). Through these computational models, the users' trajectories of expression are reorganized to suit the logic of the algorithm. Facebook's "reactions" of "like," "wow," "haha," "love," and "angry" can be easily rendered categorizable by the site's computational frameworks in ways that allow easy recognition not only for users, but also for data processors: it is easy to demarcate and sort "angry" reactions from "love." Grammatization also applies not just to these emoji descriptors, but to more complex syntax. Take, for example, Facebook's newest status update format: Facebook users can now choose from a drop-down list of "feeling/activities" to articulate their present mood or cognitive process. Utilizing hundreds of options and sub-options, users can now inform their Facebook "friends" that they are "thinking about" "all the good times" or "doing something crazy," or that they are "drinking" "beer." Each of these actions is grammatized so as to be readable by Facebook's algorithms, enabling the platform to monitor and manage a user's preference between, say, "eating" and "drinking" or "beer" and "coffee."

It's a perhaps obvious but still interesting point that Facebook users can tell their friends "all the good times" *without* using the drop-down menu: they could simply input this as a status update. However, they are encouraged to self-compartmentalize their "feelings" and "activities" in ways that suit algorithmic protocols. Most tellingly, the drop-down of status activities includes categories such as "Oreos" and "Starbucks," which if selected, automatically link to the commercial Facebook page of that company. As such a user's status update can be reorganized to automatically display a brand as part of the post—meaning that the user's identity articulation is algorithmically grammatized to double quite literally as advertising. Such instances highlight the homogenizing influence of commercial interests on profile pages that work to redirect the modes of self-expression available to web users.

It is within these expanding yet ever more computationally rigid grammars of action that we can begin to understand the performative powers of identity articulation. Scholars such as Cheney-Lippold (2017), Cover (2012), Stone (1995), Jordan (2015), and Bassett (1997) suggest that online identity "enacts or produces" the subject that it names (Butler, 1993) and in doing so performatively constitutes that subject as such. For example, describing the personal profiles of users of SNSs, Cover states that "the establishment and maintenance of a profile is not a representation or biography but performative acts, which constitute the self and stabilise it over time" (2012, 181). Differently put, users' profile expressions—for example, status updates, Tweets, or new profile photos, do not just function as online markers of taste and preference that inform other users of what that user's identity "might be like." Rather, these identity expressions actively produce how users see each other, how they see themselves, and indeed how they are materially constituted in the world. Take, for instance, as Jordan does, times when a user's profile is "hacked." When someone posts a status update on a users' behalf, it is not only that profile that is altered, but that users' articulated self:

> That self will. . . only maintain itself and have its own characteristics if it can continue to be read and be associated with its own characteristic kinds of posts. . . . We can see this in the phenomenon of people logging into someone else's social media network and posting in ways that they would not normally post. (2015, 128)

It is in moments such as these that the performative powers of online self are revealed—our profiles do not just reflect the self but are constitutive of the self they apparently only represent. This remains the case when the "hacker" is removed from the equation: the performative qualities of online self-expression remain, even when it is the users themselves who are "in control" of their profile articulations. However, the material and performative powers of online self-expression are brought to the fore when other agents—both human and non-human—intervene in identity performance. And as aforementioned, it is the agential capacities of personalization algorithms that are the focus of this book: decision-making systems that are designed and indeed encouraged to act in the users' stead and to speak for them in ways that entangle the algorithm in identity construction. It is the agential capacities of personalization algorithms to articulate the desires, intentions, and expressions of the self that, I argue, create a struggle for autonomy between user and system—a struggle that arises in numerous participant accounts featured in the next three chapters of this book. I explore this further in Chapters 5 and 6, but here I want to stress that this book takes the performative quality of profiles not as an argument, but as a *foundational premise*: more broadly in our articulations of online self-expression, and more specifically in those self-expression that are articulated in, through, and with algorithmic protocols.

THE AUTHENTIC SELF?

The notion that profile expressions can be considered performative in many ways seems antithetical to how the individual is discursively constructed by platform providers. Unlike the potentially "fluid" identities of 1990s cyberspace, web users are at present being compelled more and more to "authentically" and "truthfully" reflect their selfhood in their profiles as a stable and coherent identity. It is a practice common on the contemporary data-tracked web: Facebook insists on a "real name" policy that users must adhere to their "offline" name as part of their profile; Twitter "verifies" the accounts of users that they believe to be "authentic"; and

Google—one of the first adopters of this move to create coherent, truthful profiles—now insists that Google users use one account to access their YouTube, Google, and email profiles.

As I explored in Chapter 2, even as contemporary data harvesters look to "dividuate" the self in the name of algorithmic personalization, they continue through the preceding mechanisms to imply that users "have one identity" (Zuckerberg, cited in Van Dijck, 2013). In doing so, Facebook reinforces an idea of the singular, authentic self, which, as I have explored, has been challenged by a plethora of scholars. Specifically in relation to online identity performance, Hearn (2017) goes so far as to argue that the drive to legitimize individuals users as "authentic" on Twitter is indicative of new forms of "ideal type" (Weber, cited in Hearn) that have emerged as a result of platforms' interventions in contemporary formations of self. Hearn argues that the "verified self" has now come to replace the "flexible personality" (2017, 74) considered as dominant in the 1990s by scholars such as Giddens and Beck. This ideal type of self still retains the entrepreneurial, self-branding, and hard-working ethos of neoliberal individualism, but is algorithmically legitimized as authentic by dominant platforms such as Facebook and Twitter. Crucially, Hearn notes that the verified self is not just algorithmically authenticated, but designed to fit the logic of late capitalism. She states:

> "Verified" not by the state, or police, or even our social networks, but by privately owned telecom and technology industries and financial institutions, spurred on to assiduously self-present by the hyper-personalized affective lures and bribes like the Twitter checkmark, we are inserted into the global flows of capital. (2017, 74)

Thus, it is not primarily the state which acts to categorize and verify the self—as is suggested in Hacking's and Foucault's early work on governance—but commercial and capitalist data economies. The research carried out by Christl and Spiekermann (2016) and O'Neil (2016) proves that this is not a theoretical mode of verification: data brokers have a substantial and powerful role in deciding who is deemed worthy of mortgages, higher education, loans, and even health care. The verified self contributes to a contemporary media life wherein the social hierarchies of pre-algorithmic categorization are reinforced rather than challenged.

Hearn's arguments carry crucial weight when considering how algorithms might relate to the performative potentialities of identity articulation online. She states that the verified self strives for authenticity, but in a data-driven environment in which the goalposts for legitimization are constantly moving. Under algorithmic "regimes of anticipation" that look to constantly value the affective production of "dataveilled" individuals, the self-brander becomes the " 'speculative' subject of big data" (2017, 73). She argues that "[t]his form of subjectivity is focused . . . on the maintenance of always malleable sets of anticipatory and liquid capacities" (2017, 73). In other words, the verified self cannot be considered inner or holistic, as Facebook and Google try to reinforce, but rather inherently iterative and relentlessly reworkable—a *recursively flexible* identity that must constantly prove its worth as a verified subject within the global flows of capitalism.

Her final point—that data trackers look to "know" web users "in all our specificity and yet simultaneously stripped of our meaningful identities" (2017, 74)—is a notion that resonates especially with the qualitative data I analyze in Chapters 4, 5, and 6. This is because it suggests force relations that are exerted on the contemporary data-tracked self and yet pull that self in opposing directions. By this I mean that the contemporary drive to personalize first demands that users perform an "authentic" and "holistic" self (a demand complicated by performativity). Cover (2016), Hearn (2017), and Van Dijck (2013) recognize this as a core part of how platforms view the user. In his work on social media profiling, Szulc calls this formation of selfhood "the anchored self" (2018, 1): a self tied to a particular profile, "authenticated" (Hearn, 2017) through that profile and commodified through it. This anchored self is a market necessity: the self must be tied to an identifiable profile under current systems of algorithmic profiling if platforms are to generate revenue via micro-targeting.

Simultaneously, though, and as Cover (2016), Jordan (2015), and Szulc (2018) note, through these same market imperatives users are simultaneously asked to endlessly express themselves, to prosume, to articulate, to record their fluctuating moods and tastes, to publicize *all* that is relevant to their current preferences, and to self-categorize their digital everyday engagements through commodifiable grammars of action. Facebook's Ad preferences profiles are a prime example of how the authenticated self is pulled apart by this expressive, recursive, and ever-changing profile—as explained in Chapter 1, the hundreds of recursive markers assigned to any one Facebook user ("feminist," "dog," "tea," "hope") change on a daily basis, shift, and modulate as they iteratively inform users' future ad categorizations. This dividuated, endlessly expressive, and recursive self can again be considered a market necessity: without endless re-categorization, Facebook would have nothing "new" to sell to its marketers. Szulc argues therefore that social media users "are performing identities which are capacious, complex and volatile but singular and coherent at the same time" (2018, 14).

Szulc recognizes that this dynamic mirrors the "push-and-pull struggle" (2018, 14) between historical formations of selfhood as fixed and inner but also a workable, neoliberal project, yet argues that the relation between the anchored and abundant selves is "not really a struggle but a generative force at the service of datafication" (2018, 14). Finally, he notes, like Gillespie (2017) and Bucher (2016), that to understand how these data-imposed force relations are "actualized, negotiated or resisted requires further empirical research focused on profile users" (2018, 14). Once again, then, a call emerges from contemporary algorithmic cultural theory to turn to the users themselves. The latter half of this book responds to these calls, and in doing so contributes to current debates on algorithmic identity by finding that the different formations of selfhood imposed by proponents of algorithmic personalization—fixed, performative, authentic, expressive, recursive—are not just monetized and made at the level of the platform, but are felt and negotiated at the level of the everyday. More than this, I argue that approaching theories of algorithmic selfhood through the testimonies of users reveals an *inherent irreconcilability* in these demands on selfhood. There is a struggle at work here: an impossible tension placed on users by platforms, and felt at the level of the everyday, as I will expand on in Chapters 4, 5, and 6. First I would like to consider another relation pivotal to algorithmic personalization in the everyday: "the anticipated user" entangled with "the user herself" (Gillespie, 2014).

THE ANTICIPATED USER VERSUS THE USER HERSELF

As introduced in Chapter 1, the performative implications of behavioral targeting, data tracking, and algorithmic profiling have resulted in what scholars have called "database subjects" (Jarrett, 2014, 27), "algorithmic identities" (Cheney-Lippold, 2017, 5), and "data doubles" (Lyon, 2014, 6). These seemingly spectral manifestations seem to both "haunt" the users that they mirror and intervene in their experiences of the web. From the back-end databases in which they inhabit, the algorithmic self and the database subject do not just reflect, but also *act on* the accompanying selves they are designed to represent.

However, as Cheney-Lippold (2017) is keen to acknowledge, these ghostly figures are not the holistic entities that the preceding names might imply. Despite Facebook and Google's insistence that data tracking exists to improve *your* personal experience, there is no singular "self" that mirrors each web user: there are only abstract, dividuated constellations of data that disassemble the social subjects in ways that can be translated back into recognizable figures, as detailed in Chapter 2. Whether apparition or constellation, these dividuated algorithmic mechanics certainly *do* reconstitute the ways in which web users encounter digital interfaces, content streams, and services. There are stark and material consequences for the web users tracked and algorithmically anticipated by first- and third-party companies online: the work of Skeggs (2017) most recently reveals the potential implications in being relentlessly tracked, datafied, and valued. Skeggs notes that Facebook's tracking and profiling of users works to recreate and reinforce existing inequalities of class and wealth, potentially creating a kind of data "underclass" that marketers and businesses can either discriminate against as a credit risk or exploit in high-interest loan lending. From the perspective of the users, the valuation of their "worth" also has implications for who they interact with on Facebook: Skeggs (2017) suggests that people who are deemed of more commercial worth by the platform are more likely to be visible to their "friends" on their News Feeds. Cheney-Lippold (2017), Chun (2015), Pariser (2011), and O'Neil (2016) echo these claims: for example, O'Neil charts the extensive efforts of US colleges to ensure that the "relevant" people are targeted with high- or low-debt education. These algorithmic systems do not just match the "right" products with the "right" consumers, as proponents of algorithmic personalization often claim—they discriminate, govern, and regulate the users deemed to be of less worth to marketers and other interested parties in ways that perpetuate existing gender, class, and race hierarchies.

These scholars offer us much-needed cause for alarm against monolithic data brokers that at present enjoy a huge amount of legal and social power in setting the terms by which users are deemed algorithmically "legitimate" or "illegitimate." Such critiques take on an even more profound importance in light of the legislative deregulation of data markets. For example, the 2018 implementation of the EU Open Banking Law now gives non-financial bodies the power to access, monitor, and manage individual bank accounts and income (Open Banking Europe, 2018). Contrary to claims that Open Banking simply facilitates "more consumer choice"(Open Banking Europe, 2018) Skeggs (2017) argues that access to bank details adds yet another way for data brokers to discriminate against low-income users who are considered lucrative targets for high-interest loan-lending.

In the midst of these justifiably bleak critiques, I want to consider an observation made by Gillespie in regard to "algorithmic identity":

> These shadow bodies persist and proliferate through information systems, and the slippage between the *anticipated user* and *the user*

> *herself* that it represents can be either politically problematic, or political productive. (2014, 174, my emphasis)

It is the "politically productive" possibilities that emerge from the relationship between the anticipated user and the user herself that I want to briefly consider. The productive outcomes of this relationship might help to illuminate the nuances of entanglement that algorithmic personalization creates between user and system.

To turn to Cheney-Lippold's (2017) aforementioned claim that "we are data." As Cheney-Lippold stresses, our reconstitution in data is in many ways problematic for reasons such as those cited in the preceding. However, Cheney-Lippold also argues that because of the modulatory, self-referential process that constitutes the measurable type, there is a certain potential freedom —a "postidentity" politics (2017, 85)—that might be a source of liberation in being constituted by algorithms. For example, when comparing Google's categorization of "gender" to the "gender" assigned to individuals through established "offline" cultural markers, Cheney-Lippold argues that "Google's own interpretation of gender, faithful to nothing but patterns of data, can be dynamically redefined according to the latest gendered data" (2017, 9). Differently put, to be algorithmically identified as "male" when one is culturally identified as "female" (and vice versa) reconstitutes not just that users' potential possibilities for selfhood but can perhaps productively redefine these normative categories of gender expression in and of themselves. In other words, inside the database "identity is beholden to algorithmic fit, not the disciplinary confines of political identity" (Cheney-Lippold, 2017, 66). In doing so, the modulatory nature of algorithmic personalization might open up—as well as shut down—possibilities for how the self is (re)constituted in everyday life.

Addressing an entirely different set of possibilities, Jarrett's (2014a) work similarly looks to find politically productive value in user engagements with social media platforms. She refutes critiques that posit social media prosumption as wholly exploitative: such critiques argue that through the conversion of user-generated content into revenue-generating data, users become unpaid laborers upon which platform providers profit. Though Jarrett accepts that social media use is regularly commodified via platforms' data collection, social media engagement still holds inalienable and meaningful social value for the users that act as data providers. She proposes that commodification of user data is only possible because "before it becomes user and (usable) data, the 'like' is first a manifestation of an (already existing) set of social affinities, affective interactions or personal desires that satisfy nonmaterial need" (2014a, 20). Jarrett argues that "the affective intensity associated with exchanges on Facebook does not lose its capacity to build and sustain rich social formations even if, later, it enters into the commodity circuit" (2014a, 20).

Data provision can and should be considered as exploited by platforms such as Facebook through their commodification of it: there is undoubtedly an asymmetrical power relation in the data-for-services exchange upon which algorithmic personalization relies, one that favors the platform over the user. However, like Jarrett, as well as others such as Cohn (2019), I argue that to call social media use *only* exploitative is misguided. As the participants accounts that structure the next few chapters highlight, web users undertake complex and nuanced engagements in their positions as data providers that suggest they are far from unwitting "dupes" in their data provision. Some know they are commodified, many reluctantly agree in whole or in part to commodification, and indeed some find a form of validation in the data tracking which they permit, as I will explore.

The productive possibilities of engagement with personalization algorithms lead me thus to propose that it is necessary to investigate not just "algorithmic identity" but the selves who are entangled within the algorithm. We must explore both the anticipated users and the users themselves, the dividual and the individual, and the self as constituted both inside and outside of algorithmic anticipation in order to understand the nuance of how algorithms intervene in everyday life. To do so, I conclude this chapter by turning to the ways in which these understandings might be achieved—how social subjects themselves conceive, construct, and indeed imagine "the algorithm."

ALGORITHMIC IMAGINATION AND THE ALGORITHMIC IMAGINARY

How is do social subjects "know" the algorithms that increasingly seek to "know" them? As I stipulated in Chapter 2, algorithmic personalization creates a myriad of epistemic uncertainties for the web users who encounter it—we cannot and do not know the specificities of when, how, and indeed why algorithms intervene in our lived digital trajectories. Scholars such as Finn (2017), Chun (2015), and Brunton and Coleman (2014) seek to illuminate the workings of algorithms not as material objects, but as the abstract concepts that they are: after all, an algorithm is not a piece of "hardware" or even "software"—it is a set of instructions given to computers to execute a task that can be performed on a piece of paper as much as it can be with a computer. Finn explores the difficulties of pinpointing what exactly algorithms are: often imbued with a kind of "magical" power, algorithms seem to be "pieces of quotidian technical magic that we entrust with booking vacations, suggesting potential mates, evaluating standardized test essays, and performing many other kinds of cultural work" (2017, 16). He argues that, far from being self-contained conceptual objects, the "figure of the algorithm" is a "deliberately featureless, chameleon-like pass-through between computational and cultural logics" (2017, 52). For Finn, algorithms are "cultural machines"—"complex assemblages of abstractions, processes, and people" (2017, 2) that have come to play a messy and undefined yet pivotal role in human understanding of knowledge, narrative, and creativity.

Finn introduces the notion of the "algorithmic imagination" (2017, 135) to describe the complex, abstract, and in many ways unreachable (from a human perspective) forms of creativity produced by adaptive, deep learning computational systems. For Finn, key to a human understanding of algorithms is to undertake an "algorithmic reading" of the machines on which cultures increasingly rely: namely, to use the language, grammar, and protocols of algorithms themselves to demystify machine learning. By employing algorithmic reading, social subjects can better understand the creative outputs that humans and algorithms co-construct. In doing so, the computational is demystified as "a god to be worshipped" and instead becomes a "new player, collaborator, and interlocuter in our cultural games" (2017, 192). Finn is among other scholars who call for human interactions that look *toward* the inner workings of the computational to better understand the implications that algorithmic systems have for everyday life. Brunton and Coleman argue that in order to understand exactly how the computational constitutes the world, we must "get closer to the metal" when it comes to hardware *and* software, not in the sense of "materiality" but through interrogation of the "dynamics of the machine" (2014, 97), to understand what we ontologically take to be the "object" that is the computer.

I concur with these scholars that algorithms have found an almost god-like place in everyday engagement. The ubiquity and opacity of algorithmic personalization systems mean that their workings and effects are almost wholly obscured in ways that imbue predictive algorithms with a kind of omnipotent and omnipresent power. Throughout the next chapters I explore the epistemic trust—and indeed the epistemic anxieties—that the mythic status of algorithmic personalization creates. Here, however, I want to register some caution about the fruitfulness of epistemologically "peering into" the machine to better understand the algorithm.

Broadly speaking, Chun's (2008) work interrogates some of the issues in trying to demystify the computational *through* the computer. She notes, for example, that source code "supposedly enables pure understanding and freedom" (2008, 312)—when in reality its "effectiveness depends on a whole imagined network of machines and humans" (2008, 299). More specifically in regard to methodological approaches, so-called digital methods are increasingly being taken up by researchers as a catch-all form of knowledge production. Boyd and Crawford note that research based on the analysis of mass data sets are increasingly being used to produce "new . . . methods of knowing" (2011, 3), such as the analysis of Twitter data for understanding the outcomes of election campaigns. Digital methodologies have similarly been employed to shed light on the effects of algorithmic personalization systems, through quantitative "Big Data" projects such as research on the news consumption and the filter bubble, as I explored in Chapter 2.

Though Big Data analysis undoubtedly provides some useful insight into understanding algorithmic identity, it comes, I would argue, with a caveat of some substantial limitations. It is now well documented that the "back-end" or "black-boxed" specificities of algorithmic personalization technologies are extremely difficult to access. This is not just due to a lack of tools, but also to "the logic of commerce" (Chan, 2015, 1078), which limits academic engagement with some of the algorithmic tools that researchers do have access to. Take, for example, APIs[2] offered by platform providers such as Google or Facebook. As Lomborg and Bechmann note, "the usefulness of APIs for researchers is very much dependent on the developers and commercial providers of the service" (2014, 260), who can "freely decide" to impose whatever restrictions they require to commercially protect their data set. Thus to utilize such tools often means adhering to the commodifying grammars of action that in many ways constitute the very structures that warrant critique.

I stated in Chapter 1 that this book is interested in "situated subjectivity" (Haraway, 1988) as an approach to knowledge production. Though feminist researchers have used this term more broadly to stress that knowledge is always situated in partial, context-specific, and located perspectives, the notion of the situated subject seems especially apt in relation to data tracking. As Manovich (2011), among others, has highlighted, the web users who provide data for platforms have at best only partial insight into what happens to their data once it has been harvested (a point raised by many of this book's participants). The limited insight afforded to data providers highlights for me the situatedness of the users who perhaps benefit from but who are not strategically empowered by algorithmic personalization. I explore this further in Chapter 4, but here I want to stress that it might be tempting to overlook the situated subjectivity of data providers as less valuable, perhaps "more biased" than the large-scale, seemingly generalizable, and macrocosmic forms of knowledge production that Big Data apparently facilitates. For example, Ruppert et al. (2013) suggest that Big Data research methods allow researchers to circumvent the "self-eliciting" bias of qualitative data collection because "digital devices are modes of observation that trace and track doings. In the context of people, instead of tracking a subject that is reflexive and self-eliciting, they track the *doing subject*" (2013, 35).

Yet, from the standpoint of situated subjectivity, it is the reflexive and self-eliciting qualities of knowledge that are viewed not only as unavoidable, but thoroughly productive. I would argue that though data tracking might attempt to know the "doing subject," Ruppert et al.'s own analysis actually problematizes how and why the dividual (*not* necessarily the subject) is constituted as a "doing" entity. As outlined in Chapter 2, the methods employed by data trackers are often not interested in tracking individuals as "people" (Ruppert et al., 2013) but rather as assemblages of nodal correlation. Such methods, to me, thus throw the pursuit of knowing the "doing *subject*" very much into question.

As boyd & Crawford claim, Big Data analysis enables a potentially "profound change at the levels of epistemology and ethics" but that this change must be reflexively criticized to combat the assumption that "the numbers speak for themselves" (boyd & Crawford, 2012, 665–666). I want to emphasize that although the numbers may not "speak for themselves," those web users who encounter personalization algorithms can and do. This does not mean assuming that data captured outside the algorithm are somehow more holistic than those of inside. Though the reports of the social subjects are always only ever partial, this is what makes them nuanced, complex, and fruitful for understanding the situated knowledge of everyday life.

Though not concerned specifically with personalization algorithms, researchers Couldry et al. argue that "equally important [in regard to Big Data analysis] is the study of how social actors *themselves* deal with the increasing embedding of quantification, measurement and calculation in their everyday lives and practices" (2016, 3). Furthermore, the works of Turkle (2011), boyd (2014), Skeggs and Yuill (2016), Bucher (2016), Lapenta and Jørgensen (2015), Best and Tozer (2012), and Kennedy et al. (2015) incorporate qualitative analysis of technology users as reflexive subjects. Increasingly, projects such as Skeggs and Yuill's (2016) aforementioned research and the Oxford Internet Institute study on "Learning and Interaction in MOOCs" (Gillani and Eynon, 2014) are further reinforcing the value of collecting "small-scale," user-articulated data alongside Big Data sets in order to build a more robust epistemological picture of how online environments are used by individuals. The agential capacities of users as reflexive subjects give them a voice outside of and alongside the algorithm, and this voice allows researchers to ask and answer a different set of questions from those facilitated by Big Data analysis.

Analyzing algorithmic engagement specifically, Bucher's (2016) own work draws on the testimonies of users themselves to understand how they encounter algorithms in their everyday online engagement. Not to be confused with Finn's (2017) "algorithmic imagination," Bucher employs the term "algorithmic imaginary" to describe the ways in which users themselves understand, interpret, and engage with the algorithms they encounter on a day-to-day basis but do not "know" the inner mechanics of. Drawing on interviews with Facebook users, she explores the ways in those users negotiate Facebook's attempts to algorithmically intervene in and reorder their News Feeds, as well as show them personalized ads and suggested content. Her findings illuminate that users deliberate with, affectively connect with, play with, and work against the algorithm in order to get what they want from it. She proposes therefore that "what the algorithm does is not necessarily 'in' the algorithm as such" (2016, 40) but rather is constituted partly through the imaginaries of the users that encounter the computational. However, Bucher (2016) stresses that there is nothing fictional about the "force relations" that algorithms create: the interventions of algorithms in users' informational and content streams have performative effects on how those streams are presented, and by extension how users engage with their own sense of self.

Bucher's argument that the performative power of algorithms emerges not just from within the computational but from users themselves is pivotal to the motivations of this book. I would argue, too, that by exploring the "algorithmic imaginary" as a productive avenue for understanding and critique, we can also start to understand the complex and co-constitutional relationship between the anticipated users and the users themselves. As Bucher argues, "as algorithms are becoming a ubiquitous part of contemporary life, understanding the affective dimensions—of how it makes people feel—seems crucial" (2017, 42). In understanding how users themselves feel and negotiate algorithmic personalization, we can also begin to understand how the self takes shape within the context of contemporary late-capitalistic cultures.

More than this, though, as Bucher herself acknowledges, the ways in which users understand and negotiate algorithms have an iterative and recursive quality to them: our engagements with algorithms affect the ways in which algorithms operate. As such, taking seriously the accounts of users should not be considered some kind of interesting "addition" in understanding algorithmic interventions into the everyday. Their accounts are a vital part of understanding how algorithmic governance does and does not operate in ways that avoid overstating the disciplinary power of the computational to determine lived experience. As this book hopes to highlight, there is great value in analyzing the accounts of these so-called doing subjects—and they produce new insights into the ways in which users are "learning to cope with," trusting, distrusting, being legitimized, or being disciplined by algorithmic personalization systems as they confront them in their day-to-day interactions. The testimonies that underpin the next three chapters provide fresh avenues of understanding in regard to how algorithmic personalization creates new forms of epistemic uncertainty, struggles for autonomy, and a performative understanding of the self.

4

Hiding Your "Scuzzy Bits"

Knowledge + Control = Privacy.

—Ghostery (2014)

The fact that people monitor everything you do takes away the ability to be you in a sense . . . you can no longer choose to present yourself to the world, because you can't hide all the scuzzy bits.

—Chris (unemployed/activist/"digital miner up the North-West Passage," UK, 2014).[1]

To "do" identity in the age of ubiquitous data tracking is to be entangled in algorithmic personalization systems that, under the guise of apparently providing web users with a "personalized experience," seek to intervene in users' web engagements, to know users' identities, to anticipate their everyday preferences, and indeed even co-constitute their sense of self. Though there are always human actors at work in identity constitution, these actors are increasingly joined by algorithmic agents; thus the possibilities of personhood are structured through, in, and with the computational. Yet, to be computationally constituted does not mean that we can *only* be understood as entirely algorithmic entities: it is an obvious yet important point that life outside the algorithm goes on.

As I argue in the previous chapter, the persistent presence of personalization algorithms in users' everyday web engagements creates new formations

Making It Personal. Tanya Kant, Oxford University Press (2020). © Oxford University Press.
DOI: 10.1093/oso/9780190905088.001.0001

between what Gillespie calls the computationally constituted "anticipated user" and "the user herself" as a social subject (2014, 174). This begs the question: How do users understand this relationship? How do they negotiate their entanglement with and within the algorithmic systems that seek to "know" them? This is the first of three chapters to focus on a specific site of investigation to scrutinize the horizon of possibilities that algorithmic personalization creates for users at the level of everyday life. The platform that provides the site of investigation for this chapter is Ghostery: a browser extension and privacy tool that allows web users to view and block the commercial data trackers that harvest and monetize the personal data that web users produce as they surf the web. Taking Ghostery's former tagline and rhetorical sum of "Knowledge + Control = Privacy" (Ghostery, 2014)[2] as a starting point for thematic analysis, this chapter draws on interviews with a selection of Ghostery users in order to explore their struggles for personal privacy in the context of algorithmic personalization.

I stated in Chapter 1 that, though a related issue, this book is not primarily concerned with current debates surrounding online privacy. It might seem counterproductive therefore to orient this first qualitative investigation around Ghostery—a piece of software so clearly marked and marketed as a "privacy tool." In some ways, turning first to a privacy tool comes from the need to acknowledge that understanding algorithmic personalization necessitates not a rejection of the privacy debate, but instead investigations that go beyond it. As Cheney-Lippold notes, notions of "privacy" are changed dramatically by digital platforms' persistent attempt to "know" their users in and through data—he proposes that "the terms of privacy in a datafied world are the terms of how our data is made useful" (2017, 209). Cheney-Lippold therefore asserts that it is important to recognize that privacy is not "dead"—as popular commenters such as Cashmore (2009) and Preston (2014) have proclaimed. Rather, the ways in which privacy is framed as an individualistic pursuit hold less and less currency when compared to the highly de-personalizing ways in which the web user, as I explored in Chapter 2, is dissembled and dividualized by data tracking.

The reasons for choosing Ghostery are twofold: first, as I explore in the following, it offers a high degree of customization as part of its functionality, therefore affording a high degree of autonomy to users in deciding for themselves how to "personalize" this software. More fundamentally, Ghostery's relation to algorithmic personalization is found in its driving purpose: to reveal to users the data-tracking companies that harvest user data in order to target them at some point in their lived trajectories with some kind of tailored, predominantly commercialized, content, service, or marketing. Crucially, Ghostery also allows web users to (at least partially) block these data trackers and prevent them from collecting personal data. Many of the data trackers that Ghostery identifies specifically cite the personalization of user content and services as the primary reason for the tracking of users. For example, "commerce marketing ecosystem" Criteo claims to "make sense of digital user behaviour—across any audience—to deliver relevant, personalized ads" by predicting user interests and inferring a "deep understanding of each shopper's individual habits" (Criteo, 2018). Website optimization platform Optimizely tells its clients that "[b]eing personal is no longer optional" and promises "a complete picture of your customer that you can use to power personalized experiences" (Optimizely, 2018). Credit scorer Experian frames the imperative to "know" the user as something demanded by users themselves—they claim that "no matter the channel or device, consumers expect and demand a personalized experience from brands" (Experian, 2018). In order to meet this supposed user demand, Experian offers its clients access to customer profiles that construct a "single, persistent ID that you can use to deliver customized offers and content . . . and develop campaign reports that accurately link online and offline actions." The objectives of these commercial data trackers exemplify the argument introduced in Chapter 2 that, despite platform providers' assurances that they at least partially anonymize the user data they collect, third-party companies can and do create individual profiles anchored to particular users. They also reflect the far-reaching drive to personalize advertising, content, services, and interfaces that currently underpins a hugely successful but largely opaque tracking industry.

Though research by Turow et al. (2015) and Ofcom (2019) suggests that in the United States and the United Kingdom, web users' resistance toward commercial data tracking remains muted, web users who are interested in protecting their privacy can use a number of tracker-blocking applications that are publicly available. Web browser extension Ghostery is one such piece of software.[3] Formerly claiming to function as "a window to the invisible web" (Ghostery, 2014), Ghostery is a cost-free privacy tool that displays and blocks the third-party trackers that monitor users' site-to-site (and indeed page-to-page) movements as they traverse the web. Ghostery monitors the movements of over 2,000 advertising, data analytics, and social media trackers and indicates to a user how and when trackers are harvesting data from the user's online trajectory. Through the add-on, users can choose to block as many trackers as they wish, or conversely, can block none but still be notified when a tracker is identified by Ghostery. Its software also provides information on each tracking company should the user wish to access it, as well as details of the tracker URLs that the add-on identifies as associated with a particular page and tracker. For example, my own Ghostery add-on informs me that "54 trackers" are present when I visit *The Guardian*'s home page, the majority of which are categorized by Ghostery as "advertising" trackers, including companies such as Rubicon, Google Ad Sense, and Appnexus. Despite attracting criticism for formerly having been owned by commercial "privacy compliance" company Evidon (Simonite, 2013), Ghostery now claims to have over seven million monthly active users worldwide (Ghostery, 2018).

By allowing users to view and block trackers, tracker-blocking tools such as Ghostery seek to make explicit web users' position as "data providers" (Van Dijck, 2013) to commercial online platforms. This position is especially significant in the context of this study because tracker-blocker users can be considered to emphatically reject the subject-position of data providers—implicitly through their use of tracker blockers and, for the users I interviewed, explicitly through their testimonies, understandings, and reported experiences. However, even as this position is made explicit, it is simultaneously also called into question—by the prevention of trackers from accessing users' data. This raises some interesting questions, especially in relation to personalization: What do these privacy-concerned users "know" about the data trackers that seek to "know" them? How does their unwilling position as data providers intersect with their understandings of privacy, algorithmic anticipation, and identity? Finally, what role does algorithmic personalization—the very practice that underpins the data tracking they seek to resist—play in Ghostery users' negotiations?

The discussions, testimonies, and insights provided by the twelve participants[4] interviewed for this study offer a range of responses to these questions. Participants were recruited via calls for participation for web users who use tracker blockers on Twitter and Facebook (one of which was retweeted by Ghostery itself), as well as calls for participation posted around London and English South-Eastern digital arts and privacy organizations.[5] Based in the United Kingdom, the United States, Canada, France, and other European countries, all interviews were semi-structured and conducted via Skype, face to face, or if requested by the participant, via an email discussion. As covered in Chapter 1, all individuals interviewed for this research have been assigned or have self-selected pseudonyms in order to protect their identities. Perhaps given their interest in privacy, some interviewees—such as the self-named Participant—declined to disclose any demographic information about themselves as part of the interview process. This non-disclosure is in itself a valuable insight—suggesting that some of the web users interviewed here refuse to take up the position of "data provider" in *any* circumstances, even for research purposes.

Thematic analysis—"a method for identifying themes and patterns of meaning across a dataset in relation to a research question" (Braun and Clarke, 2013, 174)—was employed as primary mode of making sense of the data generated from these interviews. Given that themes were structured around Ghostery's own assertion that "knowledge + control = privacy," my approach can be considered as theoretical thematic analysis (Braun and Clarke, 2013, 174), in that Ghostery's preexisting discursive framework underpins the themes used to semi-structure the interviews. These themes were then combined with my overarching aims to interrogate the intersection of algorithmic personalization with identity constitution in ways that, as I outlined in Chapter 1, facilitate the critique of personalization algorithms through the qualitative exploration of everyday digital engagement. The following sections present my analysis and findings of participant testimonies—privacy-concerned web users who encounter algorithmic personalization within the context of datafication, identity, and knowledge production.

"CONTROL": TRACKER BLOCKING AS A TOOL FOR AUTONOMY

Though the Ghostery add-on and its accompanying website have (like many software applications) undergone a number of changes in the past few years, the idea that the add-on can help users "take back control" of their data has largely remained a consistent selling point. This claim to greater control is perhaps unsurprising: like many tracker blockers, Ghostery exists in response to commercial data aggregation that, as I charted in Chapter 2, have historically afforded little control to web users in regard to how tracking technologies have been used to collect personal data from web users' browsing devices. It is understandable, then, that for almost all of the Ghostery users I interviewed, the lack of control over their online and personal data emerged as a central topic of concern.

For Yoda (IT user support officer, UK), Gyrogearsloose (unemployed, Canada), Katherine (managing director, Netherlands), and Mary (web developer, US), Ghostery was at least initially framed in our discussions as a valuable opportunity to intervene and respond to the data tracking involuntarily imposed on these web users. For example, when I asked Mary why she used Ghostery, she replied:

> I use quite a few add-ons to try and prevent tracking and things . . . you know, [Ghostery] was just another way to have more control over what was going on in the browser . . . it sort of allows you to be a little bit more in control of things.

She added that she felt she "had absolutely no control over this data." Similarly, Gyrogearsloose's interview started with a very positive description of Ghostery concerning the high degree of autonomy he perceived it facilitated: "I love it, it shows me what it is that's tracking . . . and it allows me to turn it off and on at will." Yoda stated about his Ghostery use that "it's all about control and who has that control." Katherine, as part of her explanation of her motivations for downloading Ghostery, told me:

> You want to be in control of who accesses whatever you are doing on the internet, and currently you aren't so you don't know exactly what will happen with the data, especially when firms like Ad Track [one of the trackers Ghostery flags] start to sell it to merchants or try to amass huge amounts of data about you.

These responses suggest that Ghostery is fulfilling its pledge to give users more control: the app seems to be giving these participants a chance to intervene in and indeed prevent the personal data tracking that they wish to resist. This emphasis on "taking back control" is indicative of the struggle for autonomy that, as I argued in Chapter 2, emerges between web user and personalization system as a result of platforms' efforts to track and identify users in the name of a "personalized" web experience. For these web users, Ghostery creates a way to reassert their autonomy as individuals who have little say in how they are monitored and categorized by data-driven algorithms. However, I want to complicate the apparent redress in autonomy that tracker blockers seem to afford by exploring participant acknowledgments that this sense of control might be just that: a *sense*, rather than a tangibly effective tool for taking ownership of their data trail.

"KNOWLEDGE": GHOSTERY USE AS UNEASY INSIGHT

> *Ignorance is bliss.*
>
> —Gyrogearsloose
>
> *Ghostery has given me a false sense of security.*
>
> —Claire
>
> *In reality, I have no illusion that my data trail is covered up.*
>
> —Christopher

Though "control" was discussed by the Ghostery users I interviewed in relatively straightforward terms, the "knowledge" that Ghostery potentially affords[6] its users was expressed through a more varied set of negotiations by participants. For Christopher (senior systems engineer, US) and HelloKitty (unemployed, UK), the app's capacity to render visible previously invisible third-party trackers was wholeheartedly welcomed—HelloKitty told me, "I enjoy it so much to see all these names [in Ghostery's blocked list] and they are blocked," and Christopher stated that he finds the trackers that Ghostery displays "mildly interesting; when I see them, I think 'nice try, evil marketing company.'" Such sentiments here suggest that there is a welcome affective pleasure to be found in the knowledge that these "evil" trackers have been successfully thwarted, blocked by Ghostery's software.

For others, however, Ghostery's capacity to enlighten the user was articulated as something of a double-edged sword. Edward (occupation undisclosed, France) actively stated that one of the reasons he disliked Ghostery was *because* of the insight into data tracking that it provides. When asked what he didn't like about Ghostery, he stated "the only thing I don't like is that it has forced my [*sic*] to open my eyes to what is happening and I feel overly self-conscious." Similarly, when asked what she *liked* about Ghostery, Claire (student, UK) replied:

> It gives me *a false sense of security* in terms of having data tracking . . . I like the fact that you can see, I just think it's interesting to see who's trying to harvest my information, say Facebook for example, I'm not even on Facebook and yet they're trying to steal my stuff.

In contrast, then, to Christopher and HelloKitty, who see the display of trackers by Ghostery as a welcome insight into a practice of which they were formerly unaware, Claire's response to the knowledge that Ghostery affords is more ambivalent—welcome, in that it offers a sense of "security," yet met with skepticism because this security may be false. Claire was not the only participant to suggest that the insight Ghostery provides is a mixed blessing: notably, Gyrogearsloose's initial enthusiasm for Ghostery became increasingly nuanced throughout our interview. Despite his earlier assertion that he "loves" Ghostery because he can see who is tracking him, when I asked him, "What do you know about the trackers that are tracking you?" Gyrogearsloose replied:

> I know nothing about them, er, some of them are kind of self-evident by their name but most of them I know nothing and very few do I actually get curious to see what they are, I don't know anything really about the trackers themselves. . . . *Ignorance is bliss* [laughs].

Gyrogearsloose's response suggests that Ghostery's initial ability to make "visible" the tracker blockers does *not* necessarily equate to increased "knowledge": although Ghostery does provide some information about the trackers it blocks, Gyrogearsloose claims he does not really know "anything about" the trackers, and indeed is not curious to know any more about them. However, despite his assertions that he "knows nothing" about third-party data trackers, it soon transpired that Gyrogearsloose had a general idea what trackers might do with his information. He told me when asked "what do trackers with your data?":

> Well the most obvious is ads directed you know, to the individual based on their searches and you know the sites that they visit, the rest of the information I assume is probably passed on. . . . That's probably what it is, you know compiling data on demographics and you know interest preferences and things like that.

Gygrogearsloose does indeed know *something* about them in a universal sense, but without further and specific details about how, when, and why his data trail is monitored, his trajectory as a data provider cannot be fully plotted, pinpointed, or revealed.

Gyrogearsloose's self-perception as someone "who knows nothing" regarding data tracking is a telling marker that notions of "expertise" have a role to play in resisting data trackers, as I will explore shortly, but here I want highlight that Edward, Claire, and Gyrogearsloose express an uncomfortable uncertainty over the kinds of knowledge that Ghostery can and cannot provide. Though Ghostery can render apparent the *fact* that they are being tracked, the precise trajectory of their personal data remains hidden and unknowable—and so, it seems, despite Ghostery's very functionality of making visible the data trackers, ignorance really is bliss.

In these accounts the *epistemic uncertainties* created by data tracking are brought to the fore. As I explored in Chapter 2, and as scholars such as Bodle (2014) and Brunton and Nissenbaum (2015) highlight, the asymmetrical relationship between platform providers and web users means that though first- and third-party data trackers have numerous and detailed means of "knowing" web users, users cannot and do not know the ways that they are tracked, managed, and anticipated. Participants' uncertainty over the data-tracking process is here articulated as a kind of epistemic *anxiety*—a fear that users cannot know in any meaningful sense what happens to their data once they are harvested from them, and how those data are subsequently used to algorithmically intervene in their digital lived experiences. For these Ghostery users, notions of "knowledge" of their data trail can only ever fall short against the sheer complexity and opacity of the data tracking they face, forcing them to take up ambivalent positions as data providers that can only lead to an anxious sense that "control" is but an illusion.

QUESTIONING THE POWER OF "POWER USERS"

I will return to *why* these users continued to use Ghostery despite the epistemic anxieties it creates, but first I would like to examine some responses of those participants who could be classed as "power users."[7] Popular in HCI studies of algorithmic personalization and in the computational science disciplines more generally, the term "power user" is used to demark those users who have higher levels of technological efficacy, understanding, and knowledge than the "average user" (Bhargava and Feng, 2005; Sundar and Marathe, 2010). Sundar and Marathe distinguish that "[p]ower users spend a lot of time using different gadgets and browsing the internet. . . . They may be classified as 'experts' requiring less navigational support than novices," whereas "[n]on-power users, on the other hand, lack the expertise and interest in adopting newer technologies and interface features" (2010, 305).

A number of participants seemed to comfortably fit the category of "power user"—Mary (web developer, US), Katherine (managing director, Netherlands), Yoda (postgraduate student, UK), Robkifi (machine learning researcher, UK), and Chris (unemployed/activist/"digital miner up the North-West Passage," UK) all recounted high levels of technological expertise and all made references to their extensive knowledge of commercial data tracking, often acquired through their occupation. The very use of the word "power" in the term suggests that these individuals' expertise makes them more "powerful" than other web users: their skills, interests, and insight will lead to a more robust sense of control and understanding over the technologies with which they engage. In light of this, can tracker blockers like Ghostery offer a form of knowledge that leads to more robust epistemological certainty? Can they afford more "knowledge"—and perhaps by extension more "control"—to these power users in ways not possible for non-power users?

For many of the so-called power users I interviewed, their expertise seemed to actually *increase* rather than *decrease* their sense of epistemic anxiety created by data tracking. Robkifi explains that his work has led to a high level of understanding of data tracking, yet describes the *unheimlich* feeling he experiences when he is being tracked:

> So for my work I do realize that a lot of information is traceable and that in order to say, to create a website, because of the way the web is organized, you do need to track people in order to see what they've been doing on the website. So I had a quite a good insight into seeing what sort of information you can collect and started noticing that more and more, say other entities started collecting other information while you were on the web, without ever being asked if it's OK to collect that data. So it was more a general feeling of,

> the German word is *unheimlich*, you don't feel entirely comfortable, that you would like to limit what is collected, and Ghostery seemed to be a fairly efficient way of doing that.

Robkifi's use of the work *unheimlich* suggests that although he may have a considerable amount of expertise regarding data tracking, he still feels an uncanny, "unhomelike" sense of discomfort around the practice. The use of this phrase suggests that a psychoanalytic reading of tracker-blocker engagement is quite possible; however, such a reading lies beyond the theoretical scope of this chapter.[8] What I will highlight, though, is that in describing the *feeling* of being tracked as *unheimlich*, it is possible to consider that the *epistemic* anxiety created through data tracking emerges also as an *ontological concern*: being anticipated via algorithmic personalization practices has implications for users' ontological security of self. I return to this in the concluding sections; however, to address so-called power use, Robkifi goes on to explain:

> The odd thing is that I work in this field so I'm fairly well aware of what's out there, but I don't have the feeling I'm on top of it, and I find that very, that bothers me and a tool like [Ghostery] probably gives you a false sense of security that you are on top of it.

Robkifi's engagement is thus far from straightforward—regardless of his expertise, he still retains the "false sense of security" that Claire reported earlier. Despite his knowledge and use of technologies like Ghostery to protect his data trail, Robkifi's sense of power as a data provider is very much called into question.

Another potential power user was Chris. A long-term activist, digital critical theorist, and online privacy campaigner, Chris's technical knowledge of tracker blocking far outstripped my own. Not only highly technically accomplished, Chris was also happy to share his intellectual thoughts in exchange for my understandings of privacy and identity. Despite his extensive knowledge and use of multiple tracker-blocking tools, Chris used Ghostery because it was "straightforward" and "easy to use"—though he stated also that he "obviously doesn't trust it." Like Robkifi, Chris also acknowledged the difficulties of using tracking-blocking tools to completely protect your information. He showed me the presence of Ghostery on his browser, but added that Ghostery doesn't seem to always block all the trackers that he wants to block:

> I'm looking at *The Guardian*, I can find 14 trackers. Now *The Guardian* I regularly go through and I regularly block things but it's obvious that they put different, different trackers appear, so I've blocked for example Quantcast hundreds of times so I don't know what's happening there. . . . So while I use it I don't bet my life on it.

To counter this reported unreliability, Chris told me that he uses a number of different tracker blockers rather than just Ghostery, some of which required programming skills and a high degree of computational understanding to use. Yet, despite his diverse technological engagements, Chris concluded our interview by stating that he could never completely protect himself online. When asked if he could track his own data in the ways he is tracked online, he replied:

> No. For me to track my own data trail, no. I'm fairly aware, as aware as most non-experts . . . you can't track your own data.

Invoking state surveillance systems rather than the commercial third-party trackers which Ghostery purports to block, he then stated that "if the NSA or GCHQ want to pawn you, they're going to . . . maybe there's two or three people in the world that are capable of not having that." Chris's engagement with tracker blocking, then—despite his status as a "power user" in HCI terms—reveals that he believes successful control over or even simply knowledge of his data trail is a near-impossible practice. The "knowledge" that Ghostery purports to provide to its users is seen by Chris to be incomplete, fallible, and fragmented.

Participant accounts raise two questions here: First, why doesn't more expertise lead to more control, as HCI studies in algorithmic personalization suggest is desired (Marathe and Sundar, 2010; Kang and Shin, 2016)? Second, why do these participants continue to use Ghostery if it fails to provide a better sense of epistemic security over one's data trail? To address the first question, it is useful to examine what it means to be an "expert" in data tracking within the context of contemporary algorithmic personalization practices. As scholars such as Brunton and Nissenbaum (2015), Andrejevic (2013), and Mai (2016) note, the legally ambiguous, cross-platform, and endlessly recursive nature of data tracking means that for any web user, it is currently impossible to meaningfully "know" your own data trail. Mai (2016) stresses that this incomprehensibility is partly caused by the fact that data trackers do not just collect or reorder *existing* forms of knowledge about users (such as their age, gender, or purchase history). Instead, the effort of trackers to anticipate and predict user behavior actively produces *new* epistemologies about users (such as their credit scores, risk to health insurers, or their categorization into a certain identity "type").

As Bassett et al. state, HCI science has tended to treat expertise as measurable: they argue that in this field, "to be an expert is to rise above a particular, and objectively defined level of competency" (2013, 17). However, the epistemic uncertainty inherent in data provision means that "expertise" cannot be defined as particular or even measurable when it comes to a user's own data trail: as Chris reminds us, it is impossible to know the kinds of knowledge that will be produced about the data subject by data trackers. Thus, the measurable expertise that apparently amounts to "power use" in HCI terminology is exposed as just the opposite: the knowledge demonstrated by these participants leads to a *decreased* feeling of power over their data trails in the face of ubiquitous and sprawling data tracking. Perhaps this lack of "power" felt by these so-called power users is why many of these participants described themselves as non-experts—they feel they cannot be experts in processes that they can never fully "know." Though Ghostery is able to provide some kind of "knowledge" that they are being tracked, it is a knowledge that ultimately creates for participants an epistemic anxiety that they might never *know enough* in order to protect themselves, no matter how "powerful" they might appear to be.

RESISTANCE: GIVING DATA TRACKERS AN "UP YOURS"

To return to my second question: Why do users—both so-called power and non-power—continue to use Ghostery if it only leads to epistemic anxiety? Why continue to use such a tool if, as Gyrogearsloose suggests, "ignorance is bliss"? Rather than use Ghostery as a concrete means of privacy protection, Gyrogearsloose described his use in more playful, resistant terms. He told me:

> Basically my motivation [for using Ghostery] wasn't to establish privacy so much as to make it more difficult for people who are tracking me, I mean I don't doubt that I'm still being tracked, but now there's an added degree of difficulty for Google, Facebook and the NSA and Canadian equivalent, you know, and they have to find workarounds and I don't doubt that they are doing that but as I say my motivation is *mostly an up yours gesture* [laughs].

Other participants echoed similar sentiments; as mentioned earlier, Christopher's phrase "nice try, evil marketing company" connotes a kind of pleasure in resisting unwanted data tracking. Claire also suggested that she used Ghostery, as well as other privacy measures such as registering "inaccurate" data (for example, her gender) on online profiling forms, as a form of "messing" with platforms such as Google and Facebook.

Ghostery use is thus mobilized as a kind of tactical, almost playful resistance against data tracking—a *symbolic* means of non-compliance, yet one that is painfully acknowledged as at least partially ineffective. In the face of the impossibility of total refusal to participate in data collection, or trying to "opt-out" of data tracking systems (as the BBC has reported, companies such Google and Facebook continue to track users who have chosen to opt out of data collection), Brunton and Nissenbaum (2015) advocate not tracker blocking but data *obfuscation* as a potential solution. They propose resistance to tracking can be achieved by flooding the system with mis-information: like Claire's tactics, giving platforms "wrong" information functions to render the data they collect unusable in their quest to "authentically" identify users. However, in keeping with Gyrogearsloose's negotiations, they stress that obfuscation cannot work as a "cinematic refusal" (Brunton and Nissenbaum, 2015, 56) of data tracking—instead, web users must resort to "weapons of the weak" (58) against asymmetrically stronger commercial data institutions that strategically wield so much more power and control over tracking processes than individual users. Thus, though an "up yours" might function as " 'a mere gesture' of registering our discontent" (58), it is a meaningful gesture in the context of the asymmetries of power that the current drive to personalize has created.

Asymmetries of power between institutions and individuals are not unique to data tracking: De Certeau's theorizations of "tactics" and "strategies" as relational power articulations make this clear. De Certeau (1984) uses the term "strategies" to denote the maneuvers that state institutions deploy to fix and govern social subjects. He notes that strategists enjoy

> a mastery of places through sight. The division of space makes possible a *panoptic practice* proceeding from a place whence the eye can transform foreign forces into objects that can be observed and measured, and thus control and "include" them within its scope of vision. (1984, 39)

By deploying strategies that reconfigure web users as data providers that can be categorized, profiled, sorted, and anticipated, these users become objects of a pervasive gaze from which, as participants have suggested, it is difficult (if not impossible) to escape. Conversely, Ghostery use can be framed as a "tactic." For De Certeau, tacticians do not enjoy the same powerful position as that of strategists. He notes that "[l]acking a view of the whole, limited by blindness . . . limited by the possibilities of the moment, a tactic is determined by the *absence of power*" (1984, 39). Under De Certeau's framework, tracker-blocker users can be considered to inhabit the position of tacticians. As Ghostery's website formerly claimed, the add-on facilitates a "window to the invisible web" (Ghostery, 2014), but like all windows, it can only give a "limited view" rather than the pervasive gaze that strategists deploy.

Unlike Gyrogearsloose and Claire, other participants used Ghostery as a site of tactical resistance not directly against data trackers but against *other web users*. For instance, Robkifi also alluded to the idea that though tracker-blocking use may not be particularly effective, it is still worth using, because in doing so "you are probably left alone a bit more, because there's too many, say, even more naïve users than I am that will provide the data that people are looking for." Robkifi here suggests that resisting trackers requires positioning oneself against other possible data providers, thus implying that resisting for him is an individualized tactic.

Such sentiments reveal that though tracker blocking might not offer robust epistemic insight into data trackers, what it can do is offer a way to tactically outmaneuver other users. By employing his powers of individuality, Robikifi can "sacrifice" the other users who are identified and dividuated (Deleuze, 1992) by data tracking. Such a response echoes the neoliberal underpinnings of personalization explored in Chapter 2, wherein the self-interested "networked individual" (Rainie and Wellman, 2014) is discursively placed at the heart of web interaction as an "agent of their own success" (Ringrose and Walkerdine, 2008, 227). It seems in Robkifi's case that the networked individual can become an uneasy site of resistance in the fight *against* data-driven monetization strategies that seek to identify the very same individual—but only at the expense of other web users.

Brunton and Nissenbaum (2015) offer a rationale to this individualized tactic, which might at first seem to exploit the ignorance and inaction of other users who must continue to permit data provision in order for resisters to "enjoy their freedom." These authors argue that at present, web users face impossible Terms of Service imposed by platform providers that require users to consent to data usage that has not even been enacted yet. Furthermore, the imposition of these unjust terms is permitted under the watch of inadequate legal protection (Brunton and Nissenbaum, 2015; McStay, 2012) that does not offer any meaningful form of security against trackers. In such a context, web users *must* seemingly "free-ride" on the backs of those less able or less interested in protecting their data trail if they are to find any form of resistance. Web users' responses to data tracking cannot be reasonable, because the terms set by the current web economy are in themselves too unreasonable.

Finally, in regard to resistance, I would argue that inherently bound in the epistemic anxieties created by data tracking is an epistemic *faith* that tracker blockers *might* help users to assert more autonomy over algorithmic anticipation. For instance, Gillespie argues more broadly of faith in the face of "information overload":

> We want relief from the duty of being skeptical about information we cannot ever assure for certain. These mechanisms by which we settle (if not resolve) this problem, then, are solutions we cannot merely rely on, but *must* believe in. (2014, 192)

I will return to this faith in technology in the face of algorithmic personalization in Chapter 6, but for these Ghostery users, perhaps simply "believing" in tracker blocking offers a form of resistance that, though potentially futile, provides relief against the epistemic anxieties that data tracking creates.

PRIVACY VERSUS PERSONALIZATION: THE DISCONNECT BETWEEN INVASION AND CONVENIENCE

Though Ghostery looks to reveal and block data-tracking systems pivotal to algorithmic personalization practices across the web, "personalization" is not a term mobilized on Ghostery's website or in its marketing materials. Conversely, though less and less foregrounded in the site's design,[9] the term "privacy" continues to feature as a benefit of using Ghostery. It should not come as a surprise, then, that though many participants brought "privacy" into our discussions of their own accord, most (though not all) were not immediately inclined to use terms such as "personalized services" or "personalization." That said, most respondents recognized that the data tracking they resisted was connected to some commercial personalization practices—practices such as behavioral profiling and targeted advertising that, as I argued in Chapter 2, are primarily undertaken by platform providers to generate revenue. For instance, Mary stated that trackers "are fundamentally trying to get money out of you somehow," while Robkifi stated that trackers were "profiling you" for "valuable" information, and Claire stated that

> I think [trackers] are just harvesting up everything they can . . . I think they are going to sell products and try to work out how to sell me things and they are probably part of the NSA [laughs].

Claire's comment, as her laughter suggests, was admittedly glib, yet she was not the only person to associate commercial trackers with state surveillance, as I shall explore in the next section. Only Edward (occupation undisclosed, France) suggested that data tracking might be undertaken for the benefit of the user, rather than the corporation or the state; he stated that trackers were tracking individuals "to get a better understanding of how people navigate on the site."

Some participants, such as HelloKitty, Gyrogearsloose, and Katherine, claimed that they liked or did not mind the personalized services made possible through data tracking in some circumstances. HelloKitty said of personalized advertising and content:

> I like them, I like personalized content but only in such pages as Amazon . . . [the] "Recommended for You" section has some other products I haven't seen that I didn't intend to buy but I'm just browsing, I get an extra idea of what's there according to what I like.

HelloKitty in this case trusts Amazon to suggest products to her that she likes—she described them as successfully providing her with products that are "relevant" to her interests. Similarly, Christopher stated of personalized advertising, "I think it's better than non-targeted advertising, but I do not like being bombarded by any kind of advertising."

Gyrogearsloose also stated that though he finds data tracking "infuriating," he was not necessarily against algorithmically personalized services:

> I mean I don't have an argument with personalized advertising, it's the fact that they're tracking you to gear that advertising and that's what annoys me.

There seems to be a disconnect here between Gyrogearsloose's treatment of algorithmic personalization and his attitude toward data tracking—Gyrogearsloose finds personalized advertising innocuous, but data tracking *in itself* is annoying. However, considering that the very point of commercial data tracking is driven by the ethos of personalizing user experiences (in this case through targeted advertising), then this statement seems to be something of a contradiction. The embrace of algorithmic personalization practices at the very same moment as advocating privacy protection highlights why public debates need to go beyond privacy to explicitly interrogate user anticipation. If not, this disconnect—between enjoying personalized services yet finding the data-tracking practices on which these services rely invasive—cannot be fully worked through and addressed.

To return to HelloKitty's response, though she expressed a positive attitude toward algorithmic personalization in some circumstances, Yoda, her partner who was interviewed alongside HelloKitty, took up a much more ambivalent position. Yoda first stated that algorithmically personalized content was "sometimes all right," yet quickly highlighted the problems he had with some forms of personalization. As a couple that described themselves as "marketers-to-be"—at the time of the interview they were planning to use social media to start a marketing campaign for their up-and-coming business—they recognized that personalization could be both convenient for web users as customers and yet presented a number of problems surrounding privacy, informed consent, and control. As such, Yoda did not dismiss algorithmically personalized content as inherently wrong, but did signify the importance of "boundaries":

> There's a limit, as in, as a marketer-to-be, or whatever, from the marketing perspective, I mean I would also try to get personalization and help the customer to, OK we'll provide the best environment for the customer and the experience for him or her to make a sale . . . on the other hand [marketers] need to understand there's a boundary where people just need to be left alone.

Yoda's recognition of the efficiency of algorithmic personalization is worth highlighting—there are times when the practices used to track and anticipate users are beneficial to users themselves, for example by speeding up an online purchase by storing a user's account information. However, even when personalized services *do* seem to provide relevance, efficiency, and convenience, the implications of this success can still be critically interrogated. The reductive implications of personalization on universal and collective consumption (Pariser, 2011; Kant, 2014) evidence this, as does the simple fact that though some personalization practices may benefit some "valued" web users (while they devalue others, as I argued in Chapter 3) through the commodification of user data, algorithmic personalization predominantly benefits the platform. Reflecting this, Yoda tentatively told me:

> If I'm going to go in [to a website], I'm going to search for something, and I would prefer to actually search it, instead of having it served somewhere, not because I like typing on a keyboard, it's because, why would, I just feel, I'm a freak, why would you have, why on earth would you, I know it's for my convenience at the end of the day, but the fact of the matter is I know it's not for *my* convenience, it's for *your* convenience . . . and I don't like it one bit.

Yoda's cautious, self-perceived "freakish" response toward autofill text systems[10] exposes a duality embedded in its functionality as a personalized service; that is, they not only are designed to offer efficiency to the user by algorithmically suggesting search queries, but also are beneficial for the provider in that they are designed, either explicitly or implicitly, to generate income. This might be an obvious point, but for Yoda and for a number of participants, this duality revealed an ambivalent engagement with algorithmically personalized services, especially in relation to the data tracking that all participants sought, through their very use of Ghostery, to resist. Yoda feels that he is acting irrationally when he rejects the convenience of autofill ("I'm a freak, why would you, why on earth . . ."), not because he prefers the seemingly unnecessary extra labor of "typing on a keyboard" when autofill could easily complete his search request for him, but because he believes that the autofill function is not just convenient for him, but convenient too for the platform. Yoda's problem with this is rendered apparent in the following exchange:

> YODA: Most people can't understand how big Big Data is, they just cannot understand that. And how a third party can be sold on third party.
>
> HELLOKITTY: Yeah we sell ourselves.
>
> YODA: Because in the end, we are the product.
>
> *Interviewer: It might sound like an obvious question, but you guys aren't comfortable being the product?*
>
> YODA: Well, if I were to be a product, I would've first of all liked to know about it, and second of all I would like to agree on it.

Thus, Yoda's status as a "product" problematizes his engagement with the personalized services that he knows fuel the data tracking he seeks to resist—again, notions of *knowledge* ("if I were to be a product, I would've first of all liked to know about it") and *control* through consent ("and second of all I would like to agree on it") become significant. Issues of control and consent surfaced in Mary's interview: she was careful to make the distinction between unwanted algorithmic personalization (such as that which was rendered possible by third-party data trackers) and algorithmic personalization practices to which she had explicitly consented. When talking about a form of personalization she actively enjoyed—the recommended section of the book review application GoodReads—she stated:

> [The Goodreads app] is something where I'm going out and I'm saying, I want to read books . . . it's not like they are pushing this as in sending me constant emails about it, ooo, have you considered such and such a book, I go to the website and I look at the recommendations, I am the one in control here. . . . [Yet] Mostly it feels like when [other sites] are talking about making things personalized to you, it feels like what they are doing is not a service to me, it's a service to capitalism maybe.

Similarly, Katherine went on to make a clear distinction between the forms of data tracking she feels that she has consented to, and the types she has not:

> When you sign up for a free service, that's the choice you make, you go on Google and you sign up for an account, everything you do on Google, that is being tracked and they have this huge database and they know a lot about you. But that is a choice because they provide you with a service and you decide to do that. With all these trackers, you don't know that they are there, you don't know why they are gathering that information, they don't know what they are going to do with it and that is basically what bothers me.

Such sentiments suggest that data tracking is permissible when it is accepted as part of the data-for-services exchange that, as I argued in Chapter 2, currently functions as a dominant socioeconomic contract between user and system on the contemporary web. Furthermore, within these articulations there are implied or explicit references to data commodification: these users recognize themselves as data providers that scholars such as Andrejevic (2011) and Fuchs (2010) claim are unfairly exploited through the extraction of value from their data trails. However, the value extraction acknowledged by these participants is not quite as simply summed up as unjust exploitation. These participants *know* they are being offered free and convenient services in exchange for their data; it is the lack of control and consent felt in their status as "products" that seems to cause the most ambivalence. These Ghostery users accept that their social interactions are being commodified; however, their concerns lie in their sense of control (or rather lack of it) in the commodification process.

The epistemic uncertainty inherent in data tracking again emerges as a factor in this exchange: it is the "unknown unknowns" (Brunton and Nissenbaum, 2015) of what will be done with participant data once they are rescinded that causes concern, as well as what (if anything) participants are getting in return. Since the time of these interviews, data protection laws in the United Kingdom and the European Union have been overhauled: the GDPR, launched in May 2018, was designed to give users more precise control over their data use by first and third parties. Though not articulated as such, the point of these renewed laws can be considered as an attempt by the European Union to redress the epistemic asymmetry (Brunton and Nissenbaum, 2015)—and accompanying epistemic anxiety—that comes with being a data provider. The question is then: Are such laws really awarding European web users more insight and control into data brokerage that, at present, others such as US citizens still do not enjoy? News reporting of the public's reception to the GDPR suggests that the public continues to find data tracking "overwhelming" (Kelion, 2018), meaning that the increased opportunities for user control intended by these laws has not quite translated into a tangible sense of security over one's personal data. I will revisit this in the concluding chapter, but here I argue that claims of web user ignorance over data tracking carry less and less currency: via legislation like the GDPR, as well as recent reported data misuse scandals, web users—and not just those using tracker blockers—are increasingly aware they are tracked as "data products."

PERSONALIZATION AS A THREAT TO THE SELF

Due in part to Ghostery's own framing of their add-on, "privacy" was discursively pre-established as the primary vernacular through which to discuss participants' engagements with tracker-blocking. It therefore seemed important to spend some time on the subject during interviews. However, the question of *if* privacy mattered to participants was already settled; the very act of installing Ghostery is a clear indication that to Ghostery users, online privacy matters. The questions then were the following: To *what extent* did privacy matter? *Why* was the loss of privacy caused by data tracking deemed so important to these participants? And finally, given the context of my research, *how* does privacy matter in the context of personalization?

For all participants, the third-party tracking that Ghostery exposes was unequivocally—and unsurprisingly—posited as unwelcome; data tracking was described as "disturbing" (Mary), "evil" (Christopher), "quite horrifying" (Claire), "annoying" (HelloKitty), "infuriating" (Gyrogearsloose), and "shocking" (Katherine). However, though privacy from data tracking mattered to all participants to some degree, the reasons given for why privacy mattered were more varied. When I asked the question of why privacy matters, Gyrogearsloose replied:

> The primary response to that is principle, I mean a Peeping Tom is subtle too but you don't really want somebody looking in your window.

Participant stated:

> Who in his right mind would question political privacy and secret ballot? Same goes for online privacy. . . . If I have the right to vote in secret every so many years, I demand the right to live on and offline in private.

Similarly, Katie told me that "what you do on your own computer is your own business, that's your right" and Claire framed privacy as a "right" multiple times throughout the interview. This framing clearly emphasizes privacy as a fundamental right or basic moral principle that must be upheld. Such a view is far from uncommon: as philosopher Lynch (2013) and others such as McGeveran (2015) and McStay (2017) have noted, this mindset is reflected in popular discourses surrounding online privacy and highlights the function of privacy in assuring the dignity and sovereignty of social subjects as state citizens. However, in this moral framing the question of why privacy matters remains somewhat unanswered: that is, an individual might have the right to privacy, but what exactly is at stake when considering the potential privacy invasion caused by data tracking?

In many ways this is a highly complex question that is specific to different sociocultural contexts. As McGeveran (2015) recognizes, "privacy" has different definitions depending on the sociocultural environment (a notion also acknowledged by Chris during our interview): for example, McGeveran highlights the differences in European or US interpretations of what is legally and ethically meant by the term.[11] Though an in-depth interrogation of how privacy is discursively constructed by different nation-states lies beyond the scope of this exploration of algorithmic personalization, I would like to highlight that both US and European conceptions of privacy largely are founded on the premise that "the concept of privacy is . . . intimately connected to what it is to be an autonomous person" (Lynch, 2013, n.p.). That is, privacy in relation to legal or political principles protects the rights of the individual to form a unique and autonomous selfhood (Lynch, 2013; Rose, 1991). Cheney-Lippold looks to a similar notion when he states that in the digital age the individual needs the right to "breathing space" (2017, 215). Differently put, to lose the capacity and space to think private thoughts means to lose the space to form an autonomous personhood away from the prying eyes of another. Lynch proposes that to be subject to the ubiquitous and pervasive gaze of another means to be objectified, "dehumanized," and ultimately controlled and regulated by that gaze. Specifically in relation to contemporary data-tracking practices, Lynch argues that "to the extent we risk the loss of privacy we risk, in a very real sense, the loss of our very status as subjective, autonomous persons" (2013, n.p.).

Participants such as Chris, Mary, and Robkifi echoed the sentiment that privacy is a matter of maintaining, protecting, and securing one's personhood. For example, Mary talked of the detrimental effects that "erroneous" data compiled by data trackers can have on an individual. Speaking about online behavioral profiling, Mary stated:

> I mean if some erroneous data point gets in there . . . [or] somebody has built a portfolio about you and your name resembles somebody else's name . . . or you have a job interview and somebody thinks that you stole something at some point because of some erroneous piece of information that somebody has stored about you, but you have no idea that's going to happen . . . you have no way of saying, of correcting these things.

Here, then, Mary expresses concern that data trackers might *mis*-identify and *mis*-profile her—the algorithmic identity constituted through data tracking might incorrectly represent her "true" and "correct" self. Similarly, Robkifi stated that commercial data tracking in the name of personalization has implications for his individuality. He stated of profiling by advertisers:

> You feel uncomfortable, [people] are being made predictable, only, because people with your profile use [pauses] Apple Shampoo, that doesn't mean that you want it, and if that's being pushed on you then that's not right. . . . There's a chance of your individuality being eroded I suppose.

Though Robkifi emphasized that "your individuality being eroded" was a strong way of putting it (he said, "if you want to make a thing about it"), he nonetheless suggests that algorithmically personalized product suggestions might result in his "individuality" being diminished. Finally, Chris framed privacy in a way that reflects the right to privacy as a right to form an autonomous personhood:

> There's a definition of privacy and one of them is it's the way we selectively present ourselves to the world, so of course [data tracking]

> stops that, you can no longer choose to present yourself to the world, because you can't hide all the scuzzy bits.

Chris poignantly reiterates that

> [data tracking] takes away your agency . . . the fact that people can monitor everything you do takes away your ability to be you in a sense.

There are a number of points I want to make about Chris, Robkifi, and Mary's framing of selfhood. The first is that these responses go beyond simply framing privacy as a "right"; for Chris, Mary, and Robkifi, privacy in relation to data tracking is about protecting one's personhood from being mis-profiled, eroded, or "taken away" by the ubiquitous gaze of dataveillance. Data tracking and algorithmic personalization become a kind of force from which a user's sense of self must be protected. This to me suggests that the anxieties created by algorithmic personalization are not just epistemic—in that users cannot "know" what data are being collected about them and how they are being anticipated. Rather, these anxieties emerge as an *ontological* concern—in that users' sense of self is open to destabilization or erroneous reconstitution through algorithmic personalization.

The second point I want to make relates to *how* the self is framed in relation to algorithmic personalization. The participants' sentiments that the self must be protected echo arguments that meaningful privacy entails being permitted a space for being (Cheney-Lippold, 2017), or that private thoughts should be considered as forms of personal property (McStay, 2017) that are ours to disclose or retain as we see fit. For the purpose of my argument, I want to emphasize that such framings also configure the self as an entity that *preexists* the data-tracking process and which can be managed independently from the influence of personalization. Mary's and Robikifi's words especially suggest that this preexisting self is internal and in possession of the individual in question—Mary has a "true" self that can be mis-profiled without her consent, and Robikifi's self is open to destabilization via behavioral profiling. Such framings imply that in order for data trackers to represent a threat in the form of destabilization, the self must exist in some kind of stable form *prior* to the forced disclosure triggered by the data-tracking process. This existence of an inner, preexisting, and even "true" self, as explored in Chapter 2, has its roots in Cartesian models of the self that have persisted for many centuries—as Lynch states, though many of Descartes's theories of consciousness have long been rejected, "the idea that the mind is essentially private is a central element of the concept of the self" (2013, n.p.).

This lasting idea of the self becomes especially important if we consider the role of the state as a concept in relation to data tracking. The Snowden scandal was around six months old when interviews for this study took place, and Gyrogearsloose, Participant, Claire, and Chris all mentioned state surveillance and the US NSA (National Security Agency) in their discussions of privacy. Gyrogearsloose told me that Google and the NSA were "probably on equal footing in terms of one being as bad as the other." Similarly, Claire speculated that data trackers are not only "trying to sell something," but that they were "also probably part of the NSA." Furthermore, during an exchange regarding *commercial* data trackers and their attempts to anticipate users, Chris stated in response to my question of "what's wrong with being anticipated by commercial data trackers?":

> What's wrong with being anticipated? . . . because then we go into the realm of thought crime, of people, that are being arrested or can potentially be arrested, er, for their thoughts because their thoughts are considered, they're considered to be anti the state.
>
> So due to all this discussion recently of GCHQ, etc., etc., by Snowden, there's been a redefinition of the terrorism act and now I'm not sure, don't quote me on it that it's been redefined because of Snowden but there's a clause in there that extremist behavior is anything which is a threat to the state.

The conflation of state and commercial surveillance by Gyrogearsloose, Chris, and Claire is understandable given that Snowden's revelations uncovered the extensive cooperations of commercial parties such as Google, Microsoft, Skype, and Yahoo with state forces. Snowden is invoked here as a kind of touchstone in the fight against both state *and* commercial dataveillance.

The Snowden revelations thus seem to mark a historically specific "moment" in the turn toward totalitarian surveillance; scholars such as Lovink (2016) and Seeman (2014) claim that we now live in "post-Snowden" state of surveillance. However, while the line between state and commercial dataveillance seems an increasingly blurred one, it is one that I propose needs clarifying. Take, for example, Bolin and Andersson Schwarz's (2015) theory—detailed in Chapter 2—that, unlike traditional ways of "knowing" a media audience through demographic profiling of age, gender, and so on, algorithmic profiling is more interested in "real-time," recursive, and correlational behavioral patterns, which then get *translated back* into traditional audience demographic categories. Bolin and Andersson Schwarz make explicit that traditional demographic categorisations help to maintain and perpetuate established models of state surveillance. They state:

> If the advertising industry and advertising-based tech companies such as Google have an ambivalent and complex relationship to [traditional demographic] categories, government security agencies definitely *do* care about the specific individual. (2015, 9)

To elaborate, for Bolin and Andersson Schwarz, state surveillance "works in the reverse compared to market surveillance" (2015, 9) by looking for *individualized* threats rather than *dividualized* (Deleuze, 1992) patterns of consumption. Differently put, as scholars such as Cheney-Lippold and Andrejevic stress, commercial dataveillance trackers don't "really care who you are" (Cheney-Lippold, 2017, 226): they have no interest in identifying single users, but rather aim to exploit aggregate patterns of consumption across dividuals. Consider *why* entities such as Google and Facebook seem apparently driven to breach user privacy through invasive data-tracking strategies: their goal is not ultimately to surveil individuals, but rather to maximize and influence the expressive and spending power of individuals by targeting them with personalized content and marketing.

To revisit the argument proposed in Chapter 2, due to the market drive to personalize, commercial data tracking demands the self be *not only* fixed to a specific set of identifiers, but *also* endlessly expressive and legitimizable through algorithmic "grammars of action" (Agre, 1994). Commercial data trackers are interested in the self as endlessly and recursively expressive—a self that can perpetually articulate its (potentially changing) likes, habits, preferences, and desires through computational forms of grammatization that can be commodified via user targeting and profiling.

Conversely, though state institutions exploit the dividualization strategies developed by commercial trackers, the objectives of the state in mass surveillance practices are quite different: to intervene in human action not to encourage expression or consumption, but to identify social subjects in order to discipline and prevent individualized threats. This distinction is particularly politically resonant considering claims, such as those made by NSA whistleblower Bill Binney, that the mass data aggregation made possible through collaboration with commercial trackers has actually worked to "overwhelm" US intelligence services (Whittaker, 2016) and as such has so far proved ineffective in successfully preventing planned terror attacks. According to Binney, intelligence agencies still rely on specific, focused, and individual surveillance—rather than dividualized mass dataveillance—as the most effective form of terror prevention.

This argument is presented in no way to dismiss some of the poignant and persuasive work done on the performative implications of dividualizing state citizens—for example, Bassett's (2007) work is a timely call to resist the computational categorization of social subjects under the guise of efficiency or terror prevention. However, it seems important to distinguish between state and commercial dataveillance because these different forces appear to have different end goals. Ultimately, the formations of selfhood that state and commercial dataveillance constitute and demand can be considered to be quite different. One looks to demographically profile and *prevent* the actions of the individual; the other looks to correlate the web user as a dividual in a mass data set in order to *encourage* commodificable and expressive action. It is in these differences that tensions arise in the negotiations of selfhood undertaken by web users tracked in the name personalization, as I explore in the following and in Chapter 5.

CONCLUSION: DIVIDUATED DATA-TRACKED SUBJECTS

The Cartesian model of the self as inner, preexisting, stable, and holistic has been challenged—by Foucault (1988, 1988a) (cited by participant Chris in his assertion that data tracking disciplines the subject), Butler (1988, 1989, 1990), and Stone (1997), whose work on performative identity constitution contests that the self is not inherently inner or preexisting, but is constituted by and through discursive and material frameworks. I will explore this idea of identity constitution in the next chapter, but here I want to conclude by considering Jordan's model of networked privacy in relation to social networks. Jordan (2015) notes that on social networks, the self is invoked both as an entity that preexists the network and one that is brought into existence by the network. To apply this theory to my analysis, the participants primarily invoke the *former* model of the private self—that is, these participants are protecting their "inner core" (2015, 126) from the disciplinary threat of data tracking—they frame themselves as a "complex private being that they own and parts of which they either reveal or are forced to reveal" (2015, 128). Jordan makes an important point about this conception:

> The core of a person is something that may be inconsistent, changeable and negotiated, it may be part of decentered subject, but it is still the complex inner core of a subject. Privacy in this conception is not the presumption of a self-consistent inner identity but of a complex inner identity that yet still remains each individual's to dispose of. (2015, 123)

Such a formation "presupposes that a being exists prior to being read" (Jordan, 2015, 128), rather than the self coming into existence through visibility, and here these participants frame the self as a preexistent, private formation that must be protected from the dehumanizing threat of data tracking. What was articulated as an epistemic anxiety emerges as an ontological concern: that is, uncertainty of algorithmic anticipation throws the security of selfhood into question.

In conclusion, I propose that these participants' framing of the self as preexistent, inner, and private largely corresponds with modes of state surveillance that seek to individually identify rather than dividuate the social subject. This framing of the self as an entity in which you must protect your "scuzzy bits" if you are to remain "you" somewhat ironically corresponds to Zuckerberg's rhetorical, neoliberal idea that "[y]ou have one identity" (cited in Van Dijck, 2013)—the difference being that for web users who resist data tracking, this "identity" should *not* be expressed through the network, as Zuckerberg claims. And yet this model of selfhood exists in tension with the other formations that algorithmic personalization demands; that is, a self that not only is fixable and preexistent, but also can be continuously and recursively expressed and reworked. Adding "personalization" into Ghostery's rhetorical sum of "Knowledge + Control = Privacy" reveals complexities and nuances regarding how the self is constituted by and through commercial data tracking in the name of personalization, and how the self is framed by web users interested in protecting their privacy, autonomy, and sense of self.

5

Autoposting the Self into Existence

The film you quote. The songs you have on repeat. The activities you love. Now there's a new class of social apps that let you express who you are through all the things you do.

—Facebook Timeline (2014)

I am not my app permissions.

—Sam (communications manager, UK)

The previous chapter sought to explore Ghostery users' negotiations with algorithmic personalization in relation to privacy, epistemic anxiety, and the protection of selfhood against data tracking. In this chapter I shift focus: from personalizing processes that threaten an inner self to algorithmic personalization practices that have the power to (re)write the self. To do so, the chapter examines the "autoposting" activities of Facebook's third-party apps: commercial applications such as Spotify, Candy Crush, and MapMyRun that can automatically post Facebook status updates on the user's behalf to that user's Facebook News Feed.

What is a Facebook third-party app? As Zuckerberg explained during the 2014 F8 conference,[1] third-party apps are commercial lifestyle, gaming, entertainment, and shopping applications initially intended to

Making It Personal. Tanya Kant, Oxford University Press (2020). © Oxford University Press.
DOI: 10.1093/oso/9780190905088.001.0001

be "deeply integrated" into Facebook's operational and commercial "ecosystem" (Zuckerberg et al., 2014). As Zuckerberg repeatedly emphasized during their rollout, these apps are designed to play an essential role in Facebook's ambition to become a "cross-platform platform" (Zuckerberg et al., 2014) that connects not just friends, family, and acquaintances, but the millions of platforms, websites, and services that currently constitute the web. To date, the external products and services that apps integrate into Facebook include games, movies, books and music services, ticket- and product-purchasing programs, bookmarking software, photo editors, health and fitness trackers, events managers, comic strip creators, lifestyle forums, interactive cookbooks, and stress relief programs. According to Facebook, over 80 million businesses have used their "App Center", as well as the network's developer app tools, to connect with Facebook users in order to monetize their business through Facebook (Facebook Developers, 2018). From a business-to-business perspective, Facebook's relation to third-party apps plays a substantial part of the platform's commercial imperative to monetize user data by delivering to users "micro-targeted advertising" (Solon, 2018) and by packaging user data for use by third parties. It is through partnerships with apps that Facebook also monetizes user interactions within Facebook itself via micro-payments (such as in-game purchases), and expands the company's reach by integrating popular apps (such as photo-sharing app Instagram) into Facebook's subsidiaries.

Facebook's third-party apps don't just facilitate a cross-platform "ecosystem": as the opening quote from Facebook Timeline highlights, when they were first rolled out in 2011 as part of Facebook's existing profile format (known as Timeline), apps were quite explicitly framed as a tools to facilitate user self-expression. By allowing users to "express who you are through all the things you do," third-party apps apparently enable users to articulate and *publicize* their identities by accommodating a plethora of user lifestyle preferences, "interest tokens" (Liu, 2008), sociocultural tastes, consumer choices, or even affects; "your runs, your bike rides, your cooking, your eating, your sleeping, *your happiness*, your fashion, anything you want" (Zuckerberg, 2011, my emphasis). Crucially for the purposes of this chapter, to enable these user articulations, apps were given the power to automatically post status updates *on a user's behalf.* For a number of years Facebook's third-party apps had, and some such as Spotify still have, the ability to algorithmically publish posts (herein referred to as "autoposts") to an individual's Facebook "friends" network, potentially without the individual's knowledge or immediate consent (at the time of posting). In the first few years of their existence, autoposts took on a number of forms, mostly referring to an in-app action or achievement by a user and written on behalf of a user in first or third person; for example, "xxx xxxx is listening to Serious Time by Mungo's Hi Fi on Spotify," "Batman & The Flash: Hero Run—I've just scored 22,323 points!" or "I've just run 5.99 miles on MapMyRun" (see Figure 5.1). In posting on a user's behalf, autoposting presents a valuable site of investigation in relation to personalization because in this instance what is being algorithmically "personalized" is a user's very own identity performance: the user's name and picture are mined and managed by algorithms in order to (re)present content from third-party apps as deeply relevant to, and indeed expressive of, that user's identity.

Since their initial rollout, Facebook's relationship with third-party apps has suffered from a series of data-sharing issues. In 2014 news stories emerged that Facebook was allowing apps to harvest extensive data from users' profiles, including their gender, age, hometown, political and religious preferences, and education, as well as the demographic data of their Facebook "friends" (King, 2014; BBC, 2014). In response to subsequent public concern, Facebook imposed a "lockdown" on the data collection permissible by apps, and Facebook itself promised to rollout new app privacy measures, such as "Anonymous Login" that would allow users apps access without sharing any profile or demographic data (Facebook Newsroom, 2014). Then, in the wake of the aforementioned 2018 Cambridge Analytica scandal, app approvals from Facebook were temporarily suspended altogether; in face of further data privacy concerns, the app "ecosystem" seem to be at risk of disintegrating. However, despite Zuckerberg's subsequent congressional appearance regarding Facebook's data policies, the suspension on app approval was lifted and replaced with updated security settings that promised to give users more "control"

Figure 5.1. Examples of autoposting.
Credit: Screenshots by Tanya Kant.

over app permissions. Tellingly, some of the privacy features promised by Facebook never emerged: for example, the "Anonymous Login" presented at F8 2014 never made it to public implementation. From the developer's perspective, the continued absence of such data-blocking features is not particularly surprising; given that nearly all of Facebook's third-party apps are free for users to access and enjoy, the data gained from users' profiles still function as a primary (if now more limited) form of revenue generation.

Though certain features such as the harvesting of data from "friends" are now prohibited, data mining through apps continues to persist in some form. However, there was one early "casualty" in Facebook's struggle to appease both profit-driven developers and privacy-concerned users: the practice of allowing apps to "autopost" on the user's behalf. Only months after rolling out third-party apps, it emerged that autoposting was overwhelmingly unpopular with the majority of Facebook's users, and after admitting that "people often feel surprised or confused by stories that are shared without taking an explicit action" (Facebook Newsroom, 2014), Facebook has increasingly downplayed the prevalence of autoposting by apps. The site has tightened its regulations to ensure that users are not forced to consent to autoposting as a part of the terms of use when they sign up for an app, and if they do consent they are given a high degree of control over who can see autoposts (for example, "Public," "Friends," or "Just Me"). For a time during the infancy of third-party apps, however, autoposting was a common and often unavoidable feature of Facebook app use, as the following accounts reveal.

At the time that participant interviews for this study took place (between March and June 2014), many of Facebook's autopost consent measures did not exist, meaning that autoposts were more frequent than they are now. The brief rise and fall of autoposting should not be taken as a reason to dismiss its critical significance. As the works of scholars such as Gitelman (2006) and Burns (2015) highlight, technological obsolescence seems to be an increasingly common characteristic of device and platform development: new media artifacts are framed popularly as distinctively different and inarguably more sophisticated than the "outdated" artifacts they replace. However, as Gitelman states, "new media are less points of epistemic rupture than they are socially embedded sites for the ongoing negotiation of meaning as such" (2006, 6), meaning that "dead" socio-technical practices can function as valuable sites of investigation for the ways in which technologies become entangled within the fabric of everyday life. As such, and as I will argue throughout this chapter, the very ephemerality of autoposting as a sociocultural practice warrants critical interrogation.

The ability of apps to algorithmically write in the user's stead, as well as connect *across* platforms, gives rise to a host of fresh critical questions. What does it mean for an app to automatically write an articulation of our identity on our behalf? What kind of self can be constituted and performed under this form of personalization? Finally, if we treat the articulation of selfhood on social media as a performative act—that is, acts that I detailed in Chapter 3 do not just reflect but actively *constitute* the self—then what kind of self is *constituted* not by the users themselves but by the algorithms that speak on their behalf? Drawing on the accounts of sixteen Facebook users, this chapter explores autoposting as a socio-technical practice that creates critical implications regarding performative articulations of user identity.

In total, sixteen participants were interviewed as part of this research project, all of whom took part in semi-structured, face-to-face interviews designed to explore participant engagement with third-party autoposting apps on Facebook (see Appendix for full interview list). Participants were, like the Ghostery project, recruited through calls for participation on social media sites: largely through a "Plugged-In Profiles" page on Facebook set up to recruit and publicize the project, but also through tweeted calls for participants using Facebook, Instagram, Candy Crush, and other app hashtags. Once interested participants were recruited, "snowball sampling" (Browne, 2005; Skeggs et al., 2008) was also employed—utilizing participants' existing social networks to recruit more participants. As Browne (2005) notes, snowballing can sometimes lead to "sameness" in participants—however, it also meant that I could conduct group interviews and interviews with pairs. Two group interviews were undertaken as part of the project—Kevin, Alice, Rory, and Daniel were interviewed together, as were Rebecca, Audrey, Sophie, Terry, and Steve. The participants taking part in these interviews were not strangers to each other: they were housemates, most of whom were "friends" on Facebook and therefore constituted part of each other's Facebook networks. As the following analyses makes clear, their "offline" connection as housemates added a valuable additional dimension to their interviews, in the form of dynamic exchanges between participants (as exemplified in Kevin's accounts of autoposting) that highlight how the intervention of apps into users' Facebook profiles does not just affect the users themselves—it also impacts their network.

In part due to the sampling methods employed, all participants were between twenty-four and thirty years old when recruited and were based in the United Kingdom. Once again, it is therefore important to note that the accounts of participants do not reflect the plethora of possible identities or demographics on Facebook. Their responses should not be taken as representative of Facebook users as a whole—rather, their testimonies are explored here to highlight the ways in which apps intervene in self-performances that are always already embedded in preexisting frameworks of sociocultural and economic norms, negotiations, and practices.

As the following sections explore, participants recounted a number of complex, tense, and often unwilling encounters with autoposting apps on Facebook—including apps disclosing "guilty pleasures" such as trashy songs or sexually suggestive content to participants' Facebook "friends," Spotify "adding an event" to a participant's "past," and the framing of other people's game app posts and invites as "chavvy." Participants' accounts suggest that in many instances, autoposts by apps work to intervene and at times disrupt the carefully staged identity performances that users commonly enact on Facebook (Van Dijck, 2013). By considering critical notions of "context collapse" (boyd and Marwick, 2011), "taste performance" (Bourdieu, 1984; Liu, 2008), "grammars of action" (Agre, 1994) and algorithmic capital, I will argue that apps function not just as *tools* for self-expression, but as unwanted *actors* (Latour, 2005) in the writing and performing of selfhoods on Facebook. Though such an approach means that some participant accounts are awarded a larger proportion of scrutiny than others, the ephemeral and highly context-specific nature of autoposts means that such an approach illuminates the performative particularities of how apps function as algorithmic, autonomous agents.

WHO DO YOU THINK YOU ARE? IDENTITY PERFORMANCE ON FACEBOOK

Before exploring some specific accounts of autoposting, it is first important to establish how participants constructed and maintained their Facebook personas. After all, if Facebook claims that apps help users "express who they are," who did participants think they were? What *kind* of selfhood(s) did participants seek to articulate through their Facebook profiles?

Despite Zuckerberg's much-cited claim that "you have one identity" (cited in Van Dijck, 2009), as explored in Chapter 3, the idea that the self is stable, unitary, authentic, and "whole" is complicated by theories such as Goffman's (1959) and Butler's (1988, 1989, 1990) that selfhood is constituted by and through discursive, performative, and material acts that iteratively shift over time and context. Echoing Goffman's notion of the "staged" self particularly, a number of participants articulated that their Facebook profiles reflected a shifting "type" of selfhood—rather than an "authentic," holistic, or fixed self. For example, Calum (duty manager, UK) was happy to admit that his Facebook use reflected a "version of himself"—but an exaggerated version. When asked "*Do you think your Facebook use reflects who you are?*" he replied:

> Yes and no—but maybe people see a version of me, a side of me that kind of, meta, hyper, you know, side of me. . . . It reflects an aspect of my identity.

Sam (digital communications manager, UK) also suggested that her Facebook use reflected a certain type of self, rather than an "authentic" identity. Sam seemed clear that her performance on Facebook constituted what she called a "constructed public persona" rather than a "true" self.[2] She explained what she means by her "constructed public persona":

> So it's how I want the world to see me . . . so for instance, I've had depression, and you wouldn't know that from what I said on Facebook . . . you wouldn't know if I was having a really shitty day at work for instance. [My Facebook use] is like me, it's not a completely different person, it is me, but it's not all of me. And it's yeah, it's like my *best self*.

In describing their identity in such a way, Sam and Calum highlight their awareness that their identity on Facebook is a carefully crafted performance. Van Dijck asserts that through such performances, "users . . . have become increasingly skilled at playing the game of self-promotion" (2013, 210) in ways that equate to a kind of "professionalized" display of identity. However, though this sentiment was echoed in part by some participants, it was clear that for others self-presentation on Facebook did not necessarily equate to self-promotion—professional or otherwise. For instance, participants such as Melanie (civil servant, UK), Kevin (accounts executive, UK), Calum, and Sara (customer service manager, UK) were acutely aware that they were performing a constructed self; however, this self was contingent on an acute awareness of their "invisible audience" (Sauter, 2013; McLaughlin and Vitak, 2012)—the network of friends, family, acquaintances, and even strangers that constitute their "friends" network on Facebook and could *potentially* view their performances of selfhood on Facebook. As Sauter notes, by posting to Facebook, users are "submitting themselves voluntarily to a panoptic form of constant scrutiny" (2013, 12) imposed by this audience. This "voluntary scrutiny" is complicated by algorithmic filtering imposed by Facebook on user's social interactions—that is, users do not always know exactly what status updates will appear on the News Feeds of their "friends." Added to this is the fact that many users' Facebook "friends" might expect a user's adherence to a particular identity performance dependent on their relationship with that user. Exemplifying this problem, Calum explained that

> I'm quite aware that, because I see friends who post lots of political things like all the time, or petitions all the time and you do become a bit exhausted to see that kind of stuff, um so I don't want to saturate somebody else's News Feed with things that I don't really think they're necessarily going to be interested in.

As Calum further states, he was aware that his interest in LGBT politics might not always be welcomed by his Facebook audience:

> I could easily just always go on about LGBT policies when people get bored "Oh there's Calum going on about the gay shit again and again."

Thus, for Calum, posting content to Facebook is not simply about promoting what he thinks of as his "ideal" self, it was also about not "saturating" the News Feeds of his "friends" with content that might not interest them.[3] Similarly, Melanie's performance was also contingent on the eyes of her Facebook network. She states in relation to her Facebook use that "it's about being able to be selective and thinking about who your audience is." Sara also recognized that her Facebook use was affected by the scrutiny of her network—she stated that she only posts content that she deems is acceptable to her professional colleagues, saying "I have to restrict some of my personality I suppose" in relation to the kind of content she posts, and later added that "I know I shouldn't care what people think of me, but I do." Calum, Sara, and Melanie's accounts suggest that the performed self is a selfhood at least partially constituted through the perceived desires of their invisible audience.

The idea that these participants have their audience in mind when presenting the self in arguably not unique to social media: as explored in Chapter 3, now established theories of selfhood stress that our sense of who we are emerges (at least in part) from our interrelations with social subjects. However, this still leaves the question: If the enactment of selfhood on Facebook is carefully considered and staged with an invisible audience in mind, then what role do apps play in this performance? As the following sections expand upon, the function of apps in the presentation of selfhood was revealed as complex, tense, and often unwanted—apps disrupted and intervened in these performances in ways that call their status as simply instrumental "tools" (or perhaps "props" would be a more fitting term) for self-expression very much into question.

APPS AS ACTORS: ALGORITHMIC SELF-EXPRESSION

As the previous accounts suggest, the performed self on Facebook is enacted under the gaze of "the sprawling mass of contacts most people amass on Facebook" (Marwick, 2013, 368). Kevin (accounts executive, UK)—interviewed with his house mates Alice (researcher, UK), Rory (sales manager, UK) and Daniel (graphic designer, UK)—had such an acute awareness of his invisible audience that his Facebook activity was very limited. Kevin called himself a "lurker" and explained that

> I never post anything, I never do it . . . I feel sort of self-conscious. I feel like I don't want other people to think that I'm fishing for likes or if I don't get enough likes I'm like "oh that was so embarrassing I shouldn't have put that one up" [laughs].

He explains later in the interview when asked if his Facebook profile reflects "who he is" that "I don't think [people] would really, like get very much from my profile, because I don't really contribute much." Contrary to any kind of "self-promotion" (Van Dijck, 2009), Kevin's performance on Facebook is thus very much *restricted* by a consciousness of how others might see him.

Given that Kevin's awareness of his imagined Facebook audience leads to a reluctance to perform at all, what role do Facebook apps play in such instances of limited self-performance? Along with other participants such as Beth, Sara, and Alice, it was Spotify that caused the most contention for Kevin in regard to his apps usage. The Spotify music streaming app has over 232 million active users worldwide (Statisa, 2019) and enjoys a significant connection with Facebook: Spotify users can log in or sign up to the service using their Facebook account, and can access the app through Facebook's "Apps" tags. Before connecting to Spotify via Facebook, a user must agree as part of the Terms of Service to this condition: "Spotify would like to post publicly on Facebook for you" (Facebook, 2018).

At present, Spotify users do not have to sign in to Spotify using their Facebook's accounts, and if they do choose to connect to the app there now exist a number of options for how the two platforms are connected. For example, Spotify users can choose to allow the app to autopost to a user-defined subset of the user's "invisible audience," or indeed sign in to Spotify via their Facebook account but deny permission to post their song playlists to Facebook.

However, the Spotify-Facebook connection has not always been so flexible: for a time between 2011 and 2013, *the only way of signing up to Spotify was through Facebook* (ZDNet, 2011; Spotify Community, 2013). Furthermore, at that time allowing Spotify to autopost was a non-negotiable condition of using the app. A number of participants (including Kevin, Calum, and Beth) connected their Facebook/Spotify during this period and under these conditions. Other participants could not remember why they had connected via Facebook, though told me they suspected they had to in order use Spotify.

Upon signing up to Spotify via Facebook, a user's songs have the potential to be automatically posted to their friends' via the News Feed. At the time of the interviews, Spotify's settings allowed free-account holders to listen to music as part of either a "public session"—in which a user's song choices are publicized to the user's Facebook audience, or a "private" session—in which songs are not publicized. Notably, even if set to a private session, the user's session would *switch back to a public session* "after a period of time" (Spotify Community, 2013)—which seemed to be around twenty minutes—or would switch back to a public session every time a user logs in to Spotify.

Kevin explained that though aware that his Spotify and Facebook accounts are connected, he has occasionally forgotten to switch to a "private session" on Spotify, meaning that his song preferences are then published to his Facebook friends' network. The following exchange between Alice and Kevin reveals the consequences of Spotify's autoposting of Kevin's listening choices:

> KEVIN: If you forget [to switch to a private session on Spotify] then everybody's like watching every song that you're listening to, you could be listening to complete trash [Alice and Kevin laugh] really depending on what it is, it's happened a few times to me, I didn't even realize it was posting, I feel like, loads of people liked it one time, like "what is this?"
> ALICE: And it's like Dolly Parton.
> KEVIN: Yeah it was Nickleback.
> ALICE: No way—that's so embarrassing!

Here, then, Kevin's restricted Facebook performance is undermined by the Spotify app. Even though Kevin consciously chose to limit the amount of content he posts to Facebook, Spotify autoposts his listening preferences to his Facebook network without Kevin's knowledge or consent, at least at the time of posting. In publishing his listening choices in this manner, the Spotify/Facebook connection is working very much in contradiction to Kevin's carefully performed identity on Facebook.

Not only is the app working in tension with Kevin's "lurker" performance, the app is publicizing songs that Kevin—and Alice—consider to be "trashy" and "embarrassing." Kevin and Alice's sentiments suggest that listening preferences are here considered to be "symbolic markers" of identity (Marwick, 2014, 367) or as "interest tokens" that constitute a "taste statement' " (Liu, 2008). As Marwick and Liu note, identity is in part constituted by "interest tokens" (such as songs, but also film preferences, clothing and hairstyle choice, brand affiliations, and endless other potential taste signifiers) which "serve as symbolic markers that signal something about who [users] are" (Marwick, 2014, 367). As influential theorist Bourdieu ([1979] 1989) established many decades before the advent of online social media networks, such taste statements can be considered as social classifiers of both the self and others that work to position certain (often classed) social identities as more "legitimate" than others. I revisit Bourdieu's theorizations later in the chapter in my discussion of algorithmic capital, but at this point I want to emphasize that Kevin's music choices are framed as a taste performance that partially classifies "who he is," in Kevin and Alice's eyes at least. Crucially, however, unlike the symbolic markers of selfhood traditionally established as identity-making, the songs Kevin is listening to on the Spotify app are not consciously "displayed" by Kevin as markers of taste—they in fact function as *unintended* markers that are automatically posted by the app, not by Kevin himself.

Kevin's account also exposes the role of the "like" button as a flexible signifier in identity articulation. At the time of the interviews, Facebook users could only "react" to user posts by "liking" them: other "reaction" buttons such as "angry" and "wow" had not yet been built into Facebook's interface. When Daniel later states that he makes sure his Spotify app is set to "private session" when listening to embarrassing songs, Kevin states:

> KEVIN: To be brutally honest I've done opposite, I've found a really good song and turned it off private and then played it to see who would comment [the group laughs] like five times in a row, like "I've discovered this amazing music."
>
> *Interviewer: Right, and has it ever had the desired effect? Have you ever had any likes or anything?*
>
> KEVIN: No it only gets likes when it's a terrible song.
>
> [Both laugh]
>
> *Interviewer: Is that because people actually "like" it, you think?*
>
> KEVIN: No, no it's because they're ripping the piss, I think, otherwise they're kind of like "yeah whatever, you found some music, I don't care."

The fact that Kevin believes that his "friends" only "like" songs in order to "rip the piss" reveals the complex tactics mobilized by individuals in order to *subvert* these limitations—according to Kevin, his friends are reappropriating the "like" button in order to signify their derision of Kevin's song choices. In using the "like" button to signify a form of "dislike" (or at least derision), Kevin's friends reveal what Latour calls "the risky intermediary pathways" (1999, 40) subjects follow when assigning meaning to referents. In this case, the rigid logic of positive sentiment enforced on users through the "like" button is challenged; the pathway to meaning behind the button is made slippery, playful, and ironic.

This is not the only form of subversion evident in this exchange, however—in playing an "amazing" song five times in a row, Kevin tries to present what he deems to be a publicly acceptable song to his audience. In doing so, Kevin is attempting to use the app as a "tool" to perform a revised selfhood—a self-performance constituted by the public disclosure of "amazing" rather than "trashy" songs. Unfortunately for Kevin, his efforts to take back control of his performance fall on deaf ears, so to speak—it seems that Kevin's "friends" only acknowledge his performative slippage of listening to "terrible" songs in public.

Kevin's attempted redirection of Spotify's autoposts exemplifies Gillespie's proposal that algorithmic socio-technical architectures encourage users to "orient [themselves] towards the means of distribution through which we hope to speak" (2014, 184). Kevin is attempting to be "noticed" by the Facebook/Spotify connection in the "right" kind of way (by listening to the same song five times in a row)—in order to present a socially acceptable form of selfhood, Kevin works hard to orient his actions to suit the algorithmic protocols of the two connected apps.

In Kevin's response to this publicized but unintended marker of taste, we can see how the self might begin to be disrupted by algorithmic mechanics: he believes he will be viewed differently by his invisible audience because of the app's autoposts.

More than this, though, if such algorithmic utterances are considered to be performative, then it is possible to consider how such utterances do not just disrupt but actively constitute the self. Take, for example, Jordan's (2015) argument, detailed in Chapter 3, that on social media networks the self can *only* be produced through visibility on that network. Drawing parallels with Butler (1988, 1989, 1990), he argues that the self on social media is produced and maintained via disclosure of particular characteristics—or what I would call identity markers, such as music taste, sense of humor, status update "style"—that come together to form a consistent type of self for that user. Therefore self-performance can be disrupted by other actors who intervene in making the self visible—in Jordan's theorization, by a hacker writing in that user's stead.

In Kevin's case it is not another human subject that disrupts his identity performance, but a *non-human* actor intervening on his behalf. In Latour's now well-established Actor Network Theory (2005), non-human actors in social networks are posited as having the power to constitute and affect the socio-technical interactions of other objects/subjects embedded in the same network. Barad (2007) proposes that it is not only human bodies that come to be constituted through performative actions, but non-human actors too (and indeed matter itself). In keeping with their sentiments and in line with my arguments in Chapter 2, apps can thus be considered not just as tools for self-expression, but as actors that, when performatively "entangled" (Barad, 2007) with the identities of those they seek to express, have the potential to intervene and disrupt individual identity performance on Facebook. In posting Kevin's potentially "trashy" or "embarrassing" listening preferences, the Spotify app is performing a clearly unwanted utterance of selfhood—a moment of intervention into Kevin's Facebook activity, wherein the app is revealed as a powerful algorithmic socio-technical actor. By performing an act of self-articulation on Kevin's behalf, Spotify thus reveals a power to actively (re)shape Kevin's intentional representations of identity, rather than functioning simply as a tool for self-expression. I will explore this further in the conclusion, but it is important to note that the "offending" autopost works to reconstitute the identity brought into existence via the Facebook network.

REGULATING THE SELF THROUGH SPOTIFY

Like Kevin, Beth (teacher, UK) also recounted a number of unwanted autoposts by the Spotify app. She stated:

> I didn't realize Spotify automatically shared everything [to Facebook]! It was only when someone "liked" the fact that

> I added a song to a play list and played a song that I realized. I didn't care too much, despite having a lot of guilty pleasure songs but I generally switch it to a private session now as it just seems unnecessary.

Echoing the exchange between Kevin and Alice, Beth's sentiments suggest here that music choice is a symbolic marker of taste; and by divulging her "guilty pleasures," Spotify is unwantedly intervening in her taste performance. Furthermore, Beth expressed later that she felt the autoposting of "guilty pleasure" songs could have an impact on how *others* see her on Facebook, stating that "I guess [Spotify songs] will affect how people see you, but not necessarily in a bad way."

Like Kevin's testimony, Beth's account so far highlights the power of apps such as Spotify to intervene in self-performance on Facebook. Yet Beth goes on to emphasize that apps hold even greater performative power—not only to disrupt the writing of the self on Facebook, but also to regulate and restrict the self *beyond* the boundaries of the site. As Beth explained in her interview, her Spotify app would regularly (and automatically) switch from a "private" to a "public" listening session after twenty minutes. She explained that

> half the time on my phone, if I go out and I just have my headphones on if I've left [Spotify] for a bit it goes back to the non-private setting, so um, half the time on my phone I don't do it because I've already started walking and you have to like, I don't know remember how to find [the "private" setting] or whatever.

The fact that Spotify switches from a "private" to "public" session after twenty minutes impedes Beth's capacity to comfortably remain in the realm of private listening, leading to her concern that the type of song she is listening to many not be suitable for sharing:

> I might be listening to something and then I'm like "oh I want to listen to something else" and then I'll think, I'll remember I'm online,

> maybe because I want to listen to something that's a bit more like, I don't know, that I don't want anyone to know about.

Beth explicitly states that this disclosure of her song choices is unwanted, but she does not know how to stop it, telling me that "I'd rather have it so it's a private setting all the time, but . . . I don't really know how to do that." To compensate for this lack of technological know-how, Beth came up with an alternative solution to avoid the unwanted disclosure of the songs she is not comfortable sharing:

> If I listen to a playlist quite often I'll just kind of leave it because then it will just say I've listened to that playlist rather than specific songs.

For Beth, the enforced Spotify/Facebook connection has actively led to a *restriction* of the songs that Beth feels she can listen to while she is listening on her mobile phone—to avoid the risk of publicizing a song "she doesn't want anyone to know about," Beth will only listen to specific playlists. The Spotify app's connection to Facebook thus works to regulate Beth's listening habits, redirecting Beth's self-performance through an architectural framework that encourages her to adhere to symbolic markers of music taste that she feels are publicly acceptable. The "publicness" of Spotify's listening session thus compels Beth to modify her practices to suit the perceived scrutiny of her invisible audience, despite the fact that she is enacting the very private performance of simply listening to music on her mobile phone.

Beth's tactic for coping with Spotify's autoposting capabilities exposes the power of apps not just to perform on behalf of the user but to actively redirect—and in Beth's case *regulate*—the kinds of performance that can be enacted in relation to the self. Beth's coping tactics exemplify Agre's "capture model" (1994, 110) of socio-technical organization introduced in Chapter 3. This model is implemented via algorithmic "grammars of action" that are imposed on users by computational management systems (such as Human Resource systems that monitor and reorder human activity in workplaces) and which cause individuals "to *orient their activities* towards the capture machinery and its institutional consequences" (1994, 110). In Beth's case the "capture machinery" is the algorithmic technologies employed by Facebook and Spotify, and the "institutional consequences" are the making "public" (to Beth's Facebook network, at least) of music preferences that would otherwise be private. By forcing Beth to orient her activities and regulate her music choice to adhere to a normative ideal of publicly acceptable music, the idea that Spotify simply helps Beth "express" her identity is called very much into question.

It is worth highlighting that Facebook's "capture machinery" is only able to regulate Beth's behavior here *because* of the perceived scrutiny of her invisible Facebook audience. Furthermore, the disciplinary power of the invisible audience is deployed and enforced via the logic of the "like" economy (Gerlitz and Helmond, 2013)—Beth's *private* interactions (between herself and her music player) are commodified by Spotify and Facebook, who seek to generate revenue through the making public of private listening habits. After all, it is counterproductive for Facebook and Spotify to purposely "shame" Beth through the disclosure of her "trashy" music taste to her Facebook network: commercial data harvesters look to encourage rather than discourage expressive articulations because they can later be managed and monetized. Rather, the publication of Beth's private songs to Facebook is driven by monetization strategies that look to exploit a user's taste articulations: Beth's advertised playlist functions as just that—a form of advertising, for Spotify, for the artist, and for Facebook itself. Beth's negotiations with her Spotify/Facebook connections are therefore conditioned through a complex intersection of algorithmic protocols, social surveillance, and the market-driven logic of personalization—all of which lead to a regulation of Beth's self-expressions. In structuring these conditions in a disciplinary framework, it seems that here Spotify and Facebook hold the performative power to tell Beth "who she is," rather than vice versa.

Beth's negotiations with her Spotify/Facebook connection were further complicated by another account of autoposting. She explained how, a few weeks prior to the interview, Spotify had "added an event from her past":

> [Spotify] sent me this completely random thing that came up on my phone the other day that said um, "Spotify has added an event from your past," and I was like "what is that?" and it was just that I'd listened to this completely random song like, several months ago . . . it just popped there, and it kind of annoyed me because it didn't ask me if I wanted to put it on there, it just added it on there.

Beth expanded on her reasoning for being annoyed by this unwanted addition to her "past." When asked "Was it a song that you were happy to be added?" she explained:

> It wasn't one that I minded, no it wasn't like a cheesy, it was just a random album song. . . . I felt a bit indifferent about it, about the song choice, but it felt like [Spotify] was trying to make it significant and it wasn't, because I was just listening to it as part of the album you know, it wasn't like a special thing or anything.

In this instance it is not so much the "tackiness" of the song that registers as annoying for Beth—rather, it is the fact that an "insignificant" song in Beth's listening habits has been suddenly and non-consensually demarcated as significant to Beth's "past." When asked whether she deleted the unwanted autopost, Beth replied:

> Well no, because when I actually went on to my page I couldn't see it, but then somebody liked it, so it must've been somewhere but I couldn't find where it was . . . you know when it shows just [notifications] on the iPhone, but then it wasn't like on my page or on my, it was kind of just an isolated, so I don't know where it is, or if it's still there, I don't really know.

Spotify's utterance of selfhood on Beth's behalf here takes on both an ephemeral *and* archival quality—it has been added, but Beth does not know where it is, rendering action against the offending autopost impossible. As Beth states, the song is not an identity marker deemed important enough for her to consider it as significant to her archived selfhood on Facebook, yet Beth in this instance is powerless to become editor of her written *historical* identity. Once again, the epistemic uncertainties inherent in algorithmic personalization come to the fore: Beth does not know how the post was added, who can see it, or indeed how the to delete the offending post. The ambiguous visibility of the autopost (i.e., that the post is ephemerally visible to her and may or may not have been made visible to her friends) means that she is left unsure how to combat the app's unwanted interventions.

As well as the epistemic uncertainty bound in (not) knowing if one's "past" is rewritten by an algorithm, I want to stress that Beth's Facebook/Spotify connection exemplifies the *struggle for autonomy between user and system* that, as I argued in Chapter 2, is created by algorithmic personalization. In this instance, the app's decision-making capacities to automatically "add a song" to Beth's "past" exist in direct tension to Beth's own decision that the song is indeed irrelevant to her historical identity. In doing so, Spotify's algorithmic actors work to undermine Beth's own autonomous control as a social actor, and in doing so quite literally rewrite Beth's Facebook history to suit the operational imperatives of Spotify.

In his assessment of the algorithmic self, Cheney-Lippold invokes Haraway's cyborg to blur the distinction between the self and the nonhuman other that increasingly constitutes "a subjectivity that extends into the nonhuman world" (2017, 162). The shadow of the cyborg is cast in Beth's rewritten past—a past co-constituted by her listening to music walking down the street, made public by Facebook to her invisible audience and algorithmically chronicled by an algorithmic actor purportedly acting on her behalf. This entanglement between self and algorithmic actor highlights that "our machines are disturbingly lively, and we ourselves frighteningly inert" (Haraway, cited in Cheney-Lippold, 162). Not only is there a "disturbingly lively" algorithm at work here, this algorithm is not, as Spotify and Facebook would claim, acting solely in the interests of Beth, but rather to fulfill the profit-driven operatives of these two commercial enterprises. The liveliness of Spotify's algorithmic actors are perhaps innocuous—as Beth herself states, she "does not really mind" that they are rewriting her historical identity. Yet, however mundane the intervention, the result is a subjectivity archived through and entangled in the (commercialized) algorithm: Beth, is rendered a reluctant and somewhat unremarkable cyborg, but a cyborg nonetheless.

"YOU HAVE ONE IDENTITY?" ALGORITHMIC CONTEXT COLLAPSE

Not all moments of identity performance slippage via apps exposed a "guilty pleasure" for participants. For example, Sam (digital communications manager, UK) reported that she had only experienced one instance of autoposting: by an app called Slideshare, a tool for designing, creating, and sharing professional presentations. Sam explains:

> So it turns out when I upload something to Slideshare it posts a picture of it on Facebook . . . that's why I don't like things that autopost, because I don't, I don't really use, I don't use my personal Facebook profile for works things, I use Twitter for it, so my Twitter profile is like my "work me."

Sam thus alludes to the fact that her identity performance changes depending on the platform—her Twitter account presents her "work me," while her Facebook account does not. Furthermore, as Sam herself admitted during the interview, publicizing your professional presentations does not necessarily constitute a disclosure of a "guilty pleasure." Why then was she bothered by this unintended posting of professional content? She explained:

> I think for me I guess it goes back to the like, the persona thing because I don't really talk about work on Facebook . . . it just didn't really fit with the sort of stuff I do, whereas with Twitter I'd more than happily say, in fact probably will say, this is a presentation that

I did because that's where I talk to people about work stuff I do, and I have people who follow me for work stuff.

Thus in this instance the autoposts of Slideshare do not disrupt the boundary between public and private—a boundary crossed in Beth and Kevin's negotiations with Spotify, for example—but instead a boundary between online social contexts. In dissolving the boundary between Facebook and Slideshare, the Slideshare app's actions epitomize what Marwick and boyd call "context collapse" (2011)—that is, "the theory that social technologies make it difficult to vary self-presentation based on environment or audience" (Marwick, 2014, 360). As Marwick notes, "people have developed a variety of techniques to handle context collapse" (2014, 360), and in Sam's case this entails having separate Twitter and Facebook accounts that represent different facets of Sam's selfhood. In autoposting symbolic markers of her professional selfhood to the wrong context (that is, Facebook rather than Twitter), the Slideshare app brings about a collapse between contexts that Sam has worked hard to avoid.

In creating this context collapse for Sam, the Slideshare app highlights that Facebook's ambition to become a "cross-platform platform" can tangibly disrupt the context-specific identity performances enacted by users. Such disruption works to highlight Thumin's argument that self-presentation increasingly becomes a necessary "condition of participation" (2015, 5) for web users interested in using new technologies. Szulc (2018) takes this further, arguing that in platfroms' attempts to connect users' identity performances across sites for commercial purposes, "it becomes more difficult for SNS users to create diverse self-performances" (2018, 13). We can see this process most acutely at work in the automating writing of identity across platforms and in users' stead. After all, the context collapse in this case has been caused by a work-based content creation tool that, via the logic of the " 'like' economy" (Gerlitz and Helmond, 2013) compels its users to connect to other arenas that are commonly used to perform identity articulations. In doing so, autoposts by apps highlight that users *do not* have one identity that can be "expressed" across all platforms to all audiences. Sam's sentiments exemplify Van Dijck's assertion that "each construction of self entails a strategy aimed at *performing* a social act or achieving a particular social goal" (2013, 212). The function of apps to apparently "express who you are through all the things you do" (Facebook, 2014) actually works to foreclose the possibilities of enacting multiple identities across different platforms.

More than this, though, the context collapse caused by autoposting across platforms (which is indifferent to the various presentations of self that are deemed appropriate by subjects to various contexts) makes explicit the *impossibilities* inherent in the contemporary drive to personalize: that is, to both "know" and "anticipate" the "one identity" that Zuckerberg believes exists, but whose endless expressions can be successfully captured across platforms *and* recursively used to anticipate the endlessly expressive user. As explored in Chapter 2, data trackers' tireless efforts to correlate and map the trajectories of abstract "dividuals" (Deleuze, 1992) are indicative of the fact that platform providers are not only trying to make fixable "one identity"—rather, they are trying to capture, correlate, and monetize real-time movements between dividuals as nodes in a network, creating both "abundant" and "anchored" selves as they do so (Szulc, 2018). Though Szulc proposes that these models are a "generative force" (2018) that bolster each other under the logic of commerce, I argue here that these models emerge for Sam as a struggle for autonomy that requires a negotiation of models of selfhood that are in tension. Ultimately, these formations of selfhood demand that the user, and not the platform, somehow reconcile this tension—in this case, for Sam it is working against Slideshare and Facebook to prevent context collapse. In Sam's words, "the apps I choose probably do tell people about me. But I am not my Facebook app permissions."

APP DISCLOSURE AND SEXUALLY SUGGESTIVE CONTENT

Kevin, Sam, and Beth's engagement with apps and autoposts so far revolve around the apparently innocuous leakage of "taste statements"

(Liu, 2008) that unintentionally intervene with their self-performance on Facebook and elsewhere. Calum's experiences of apps, however, involved the disclosure of more sensitive material. Calum explains:

> So what happened was, on Instagram you know, I follow all sorts of things, mostly friends but you know sometimes the occasional celebrity who's interesting on Instagram . . . but in this instance it was a porn star.

Calum explains that he was "liking" (on the photo-sharing platform Instagram) images from this porn star, some of which were sexually suggestive, and in doing so these photos were appearing as part of his Facebook activity:

> Of course these [photos] were coming up on my News Feed, which I didn't, which I wouldn't have been made aware of, only for I think another friend had actually liked it on Facebook.

Calum, like Sam, Kevin, Beth, and all other participants who had experienced unwanted autoposts, acknowledges that though he may have consented to some form of autoposting as part of the Terms of Service for using the app, he was not aware that this particular instance of autoposting was going to occur. As Calum put it, "I wasn't aware of what [Instagram] was going to be sharing . . . I understood it more as that if I took pictures and wanted to share them, they would share to Facebook"; it did not occur to him that simply "liking" a photo on Instagram would trigger an autopost to Facebook. Calum eventually figured out how to cut off the connection between Instagram and Facebook, though he admits it was "a bit of a job." Nonetheless, his experience highlights a subtle but important distinction: Calum had consented to "the app posting on his behalf" at the time of installing the app, but he felt he had not consented to the specifics of autoposting with which he had subsequently been confronted. As scholars such as Nissenbaum and Brunton (2015), Gillespie (2014) and McStay (2012, 2017) have noted, the lack of specific information and the use of opaque and vague terminology in Terms of Service mean that understanding the socio-technological conditions which users commonly accept can at times be difficult, if not impossible. It seems that for Calum merely consenting to autoposting as part of the terms and conditions of app use does not equate to unconditional consent in all circumstances.

As Calum explained, he only realized the Instagram photos were being publicized on Facebook *after* his friend had "liked" them. Such retrospective realization highlights a form of opacity unique to autoposting: unlike other posts that are consciously written by the user themselves, autoposts by apps do not always appear on a user's own Timeline or News Feed. Instead, they appear only on the News Feeds of "friends." For example, Spotify songs are not logged or archived anywhere on a user's Timeline, they only appear as ephemeral moments to their user's "invisible audience." These autoposts are invisible to the very individual who has supposedly "written" them, thus rendering action against such autoposts impossible—unless the autopost is made visible by *another user's* acknowledgment of it.

The invisibility of autoposting once again highlights the epistemic uncertainties that emerge from engagement with algorithmic personalization: users do not know for sure what content, services, or interface changes are produced because of platform's drive to personalize web users' experiences. In functioning in this epistemologically obscure manner, autoposting can be consider a kind of "technology of the self" (Foucault, 1995) that is in fact far removed from the self it supposedly speaks for. There emerges from such a scenario an almost paradoxical "entanglement" (Barad, 2007) between user and system, wherein the self is removed from the performative moment of self-constitution. To elaborate, Jordan emphasizes that the process of an individual being made visible on the network "depends on a priority from *receivers who legitimate senders* based on styles of messages and the sending of messages" (2013, 131, my emphasis). Applying this to the "material performative" (Jordan, 2013, 53) of autoposting, it seems that the self is produced by an algorithm and legitimized as a subject by receivers (in this case, the Facebook audience) *without any conscious action from the sender/subject at all.* This might explain why autoposting is so overwhelming unpopular (with both participants and the wider Facebook community)—not because autoposts are "confusing," as Facebook claims, but because they algorithmically constitute the self in ways that remove the self (at that moment) entirely from the process of constitution, even as it is brought into existence.

WANTED AUTOPOSTING?

What of those autoposts that *are* wanted by users? Marc (postgraduate student, UK) and Rory (sales manager, UK) were the only participants who said they "did not mind" if apps autoposted to their Facebook audiences. However, even these participants did not mean any app—they were specific about the apps that they did not mind autoposting on their behalf. For example, Marc, who had no specific negative experiences of autoposting, did acknowledge that he enjoyed publishing his running activities to Facebook through the Sports Tracker app—he stated that "I get a few likes now and then if it's a particularly long run." Similarly, Rory stated that he "always knows" when the Pinterest app will autopost, and "it's not necessarily that it's something that I wouldn't share anyways, it's just about the fact that you know, I'm in charge of all of this."

The key difference between these responses and Calum's seems to be the awareness enjoyed by Marc and Rory and the high degree of control that such awareness afforded them. In Marc and Rory's account, the app functions more like a tool than an actor, and thus autoposts work to support their intended identity performances, rather than disrupting or undermining them, or indeed triggering a struggle for autonomous control over what is posted and when. It seems, then, that it is the very *unwantedness* of offending autoposts that renders their function as actors apparent.

It is moments of disruption that scholars such as Latour (2005, Latour as Johnson 1988) posit as pivotal in revealing that the agential capacities of technologies do not just "support" human actions in constituting the fabric of social life. Rather, as social subjects "delegate" (Latour as Johnson, 1988) certain tasks to non-human technologies, the non-human actors in question are then afforded the power to reshape both the social interactions for which they are responsible and indeed the very human subjects who have awarded them their responsibility. Here I would argue that the algorithm tasked with aiding self-expression is delegated to with similar implications. In supporting self-expression, autoposts reconstitute how those self-expressions materialize, and though this might be more noticeable in the disruptive instance of autoposting, it applies to all the ways in which algorithms as non-human actors interject in self-performance. Algorithmic personalization can be considered a practice that, even when it *is* wanted, still reshapes the social, economic, and political interactions that constitute the everyday.

ALGORITHMIC CAPITAL: AUTOPOSTING AS "CHAVVY"

This chapter has so far centered on the interventions of apps into participants' own self-performances—yet many participants also noted the presence of apps in the Facebook activities of their "friends," which were frequently described by participants as irritating, frustrating, or annoying. As Melanie states of game posts and autoposts by her "friends":

> It's advertisements as far as I'm concerned . . . it's people I know that are advertising these things and it's crafty and I don't like it.

Melanie's observations that game posts are "crafty advertising" exists in clear tension to Facebook's rhetorical framing of these posts as the "sharing" of lifestyle pursuits that facilitate user self-expression. Her sentiments highlight the point that under the logic of the commercialized web, the affective pleasures of web users become monetized through marketing strategies that are anchored to individual profiles and connected to the profiles of others as a form of profit generation. For example, players of the popular smartphone game series Candy Crush can either pay for "tickets" to proceed to the next level of the game, or they can ask their Facebook "friends" for tickets via Facebook. Instances such as these render the value of social connectivity as profoundly apparent—the connection between three Facebook "friends" (the number of friends needed to get a ticket) was at the time of interviews quite literally worth 79 British pence (101 US cents).

Participants' framing of posts by games as "advertising" suggest that the monetary value of the social web does not go unnoticed by those users implicated in it. Posts by game apps highlight the advertising value of Facebook apps for third-party stakeholders (app developers, app owners, data aggregators, etc.) in terms of generating visibility for marketers' apps—yet these same posts, according to participants, also hold *negative* cultural value for users in relation to self-performance on Facebook. For example, Sophie (publishing assistant, UK) stated:

> My biggest reaction when I see people post gamey kinds of status things is just like I can't believe you play those stupid games, and people actually go down in my esteem.

Similarly, in their group interview with Daniel and Kevin, Rory and Alice took up the idea of game invites as "spammy" and annoying. Like Sophie, Alice and Rory believe that autoposts on Facebook affect how others see them:

> ALICE: Yeah I think I just think people are probably just a bit stupid that's really harsh isn't it. . . . I mean it's slightly hypocritical me saying this because I've clicked through terms and conditions without looking at anything, but I think it's just a sign of people not really paying attention to what they're doing, or not really having the foresight to think oh hold on maybe I should check this because games are really dodgy on Facebook.
>
> RORY: I guess there are some people who are just, so [pauses] I don't even know how to describe it.
>
> ALICE: Were you going to say chavvy?

RORY: Well, I can think of somebody who I would class as being chavvy who does, who everything comes through and you think, ah, typical.

ALICE: I hadn't thought of it as a generalization but I can immediately think of some people who would fit that bill.

Both Rory and Alice agree that inviting people to play games and posting game high scores is "chavvy"—a term that describes "young lower-class people typified by brash and loutish behavior" (OED, 2018) and that is recognized in the United Kingdom as a broadly derogatory class-based slur—hence Rory and Alice's tentative and self-conscious use of the word. For Alice and Rory, the act of even *unintentionally* allowing autoposts to bombard friends with invites is described as "chavvy." Rebecca (lecturer, UK), echoed this, stating:

> When I see people post stuff or you know sharing stuff about games, I don't think you're an idiot for playing the game, I think you're an idiot for *sharing* it.

Thus for Rebecca, it is not the playing of the game in itself that matters, it is making that game play *public* that is seen as detrimental. Here, Liu's analysis of "destructive information" in relation to the performance of the self becomes especially relevant. As Liu states:

> Any outlier of interest tokens in [user] profiles—such as the inadvertent mention of something tabooed or distasteful—could constitute destructive information and spoil the impressions that users are trying to foster. (2008, 258)

Autoposts by game apps are framed by Sophie, Rebecca, Alice, and Rory as pieces of "destructive information"—the sharing of game achievement is connoted to be distasteful or unsophisticated. These responses beg the question: Why is the practice of even unintentionally allowing game apps to autopost so "tasteless" to these participants? To answer this, it is useful to revisit Bourdieu's ([1979] 1989) notion of how individual "taste" can deployed as both an identity expression and classed form of (de)legitimization. Bourdieu notes that individuals' sense of taste—that is, "our tendency and ability to acquire (materially and symbolically) a certain class of classified and classifying objects or practices" ([1979] 1989, 173)—works to distinguish legitimate and acceptable social practices from those deemed to be illegitimate or unacceptable. Differently put, culturally agreed notions of "good" or "poor" taste expression can be deployed as a way of devaluing or valuing the social positionality of the classed, gender, and raced (among other markers) subject.

As Ignatow and Robinson (2017) note, Bourdieu's theories have gained increasing purchase in digital sociology because of their insight in explaining the formulation and distribution of economic, material, and educational "digital divides" between the technological "haves" and "have nots." Although there are numerous types of capital that individuals can deploy in order to enhance their social status, it is *cultural* capital that I am interested in here—that is what Bourdieu defines as acquired "knowledge," "manners," or "orientations/dispositions" (Jenkins, 2002, 85) that are seen to be "legitimate" by particular classes or groups and therefore "mark and maintain social boundaries" (Jenkins, 2002, 135)[4].

Nissenbaum and Shifman argue that cultural capital is acquired in digital contexts by individuals in their belonging to and acquiring of recognized "particular, desirable" canons of knowledge within certain online groups, as well as their adherence to expected social practices within that group that define and are defined by "constant normative differentiation" (2015, 194). The authors use the example of web users' knowledge of memes to legitimize their status within web forum groups; however, such forms of normative differentiation extend beyond meme knowledge to other web-based social practices: for example, the posting of "too many selfies" as a perceived unwelcome habit. To bring this back to game autoposting, participant responses of autoposting as "chavvy" and "stupid" suggest that game autoposts are classified as having low or *insufficient* cultural capital. Allowing games to autopost is seen as a breach of expected norms and practices on Facebook or as form of bad manners. As Sam puts it:

> I think the fact that they don't seem to have any self-control about sharing, so whether or not it's that the app's too tricky or forces you to invite people . . . but I think it's because I'd see it as being a little bit impolite, or it's just not my version of internet etiquette to spam people with this stuff.

There are a few points I want to make about the (low) cultural capital of autoposting. One is that is it not only suggestive of the "digital capital" that Ignatow and Robinson (2017) argue is frequently deployed (or indeed not deployed) by internet users in order to legitimate their own social status. They use the term "digital capital" (2017, 952) to explain how access to and knowledge of the web can be used to enhance users' social and educational status. Instead, I would argue that it is not digital capital that is initiated through autoposting, but in fact a kind of *algorithmic capital* that is retained or lost.

Algorithmic capital, I propose, can be considered as users' knowledge of and negotiation with algorithmic power—for example, allowing an algorithm to post game scores or "spam" friends with game invites on the users' behalf—which then acts as a classifying mechanism. It seems that for the aforementioned interview participants it is distasteful or trashy to let algorithms intervene in your self-performance; hence Rebecca's claim that it is not playing but sharing a game that makes a user an "idiot." As aforementioned, though, Bourdieu's ([1979] 1989) theorization distinguishes between different forms of capital; it is cultural capital that interests me here because the framing of other users' engagements with autoposts as distasteful exposes a value judgment made through and with a user's orientation toward the algorithm. In functioning as a value judgment, I consider algorithmic capital to function within the framework of cultural capital rather than, say, educational capital because it is not necessarily educationally-situated forms of social interaction that render algorithmic orientation legitimate or illegitimate. Similarly, though connected to social capital in that users' connections to each other might determine whether they are exposed to autoposts by "friends," I would argue that it is not a form of social capital. This is because—although social capital certainly can be strengthened or weakened through social ties with others on Facebook, as Ellison et al. (2007) establish—within the context of algorithmic capital it is not about *whom* you know but *how* you engage yourself in your algorithmic entanglements. Differently put, under the normative differentiations established within frameworks of algorithmic capital, it is "distasteful" to let an algorithm speak for you.

The fast-paced development of the digital landscape means that gaining and losing cultural capital online involves sets of practices that are often still in flux: as Ignatow and Robinson argue of memes, the struggle to accrue cultural capital on the web is often fought through engagements with "unstable cultural forms" (2016, 958). This leads me to my second point: acquiring or losing algorithmic capital is a developing socio-technological practice that is yet to be accompanied by a fixed set of norms that have defined what is a "legitimate" or "acceptable" form of automated social interaction. The work of Gillespie (2014) and Bucher (2016) is once again relevant here—Bucher's (2016) study especially illuminates how users are increasingly using their (limited) knowledge of algorithmic power in order to "game" the system, or exploit algorithmically managed forms of social interaction in order to enhance their own social needs. However, there is still much work to be done in understanding how class dynamics and notions of legitimacy might intersect with the ways in which users turn to face the algorithm. As algorithms increasingly intervene in everyday identity performance, web users' negotiations with such algorithms may become a new way to devalue each other through the demarcation of certain marginalized or "illegitimate" socio-automated entanglements.

The final point I want to make is that autoposting is an "unstable" cultural form that has implications not just in regard to algorithmic capital, but also for the performative self. Jordan notes that the changeability of communication technologies has implications for these technologies as "material performatives" (2013). He states: "what is at stake [in communicative practices] are not fixed universals but particular social and cultural practices that allow transmission to reliably and repeatedly occur"

(2013, 133). The repetition Jordan refers to here relates to Butler's proposal that it is discursive iterative acts which constitute and also destabilize the self. Butler notes that it is through repetition that subjects come to be discursively constituted, and yet it is through repetition (which is paradoxically never an exact copy of itself) that space for destabilization opens up (Butler, 1993; Jordan, 2013). As a new form of material performative, autoposting does not just carry a new means of deploying cultural capital, it also opens up new possibilities for iteration and destabilization of the self—as the accounts of Beth, Kevin, and Calum make apparent.

It is important to note that the as yet contestable cultural practices inherent in autoposting are structured not only by legitimizable social values, but also by the exchange values imposed by the "like economy" (Gerlitz and Helmond, 2013). As Jarrett notes, we must consider users "as agents who, while exercising that agency, may nevertheless be working within capital, disciplining other users into social norms and patterns of behavior that support that system" (2014a, 24). In other words, for Jarrett the commodification of users' social interactions works to potentially impose forms of discipline on those users: for instance, the disciplinary regime (partially self-) imposed on Beth as she restricts her song choices for the sake of her invisible audience. However, when it comes to game autoposts, though the classed values expressed by participants involve disciplining (through judging as illegitimate) other users' autoposting actions, autoposts seem also to be *rejected* by participants in part *because* they are viewed as a form of commodification: they are, as Sarah puts it, "crafty advertising." Thus the disciplinary power of the invisible audience seeks to regulate users through class dynamics, and yet *also* seems to reject the capitalist mechanics of autoposting as a form of advertising—autoposts are "tasteless" exactly because they are perceived to be commercial mechanisms. Autoposting therefore involves a complex negotiation of algorithmic capital, commodification, and identity articulation that suggest algorithmic personalization comes with a set of considerations that once again extend far beyond considerations of data privacy.

CONCLUSION: PERSONALIZING PERSONHOOD

My analysis of participants' accounts of (largely unwanted) moments of autoposting has sought to highlight that Facebook's third-party apps are not only *tools* that "help users express who they are," as Facebook claims—they are technological a*ctors* that hold the autonomous potential to write acts of selfhood on behalf of users. By considering the lived experiences of those who have encountered autoposting, it becomes possible to document and critically examine one of the key arguments proposed in Chapter 2: that is, personalization's attempts to act autonomously yet algorithmically on the user's behalf actually works to *undermine* the autonomous identity articulations supposedly aided by personalization.

Though this chapter has looked to examine largely unwanted moments of autoposting, is important to reiterate, as highlighted by Marc and Rory's testimonies, that not all posts by apps are viewed as detrimental to self-presentation on Facebook. Given the right level of consent, control, and understanding, apps can and are used by users to display wanted—rather than unwanted—taste articulations. The popularity of apps also suggests that many users willingly and enjoyably engage with apps on a daily basis. It thus seems it is the *unconsensual nature* of autoposts—where the app as tool becomes the app as unwanted actor that can perform and act on behalf of the user—that is resisted by participants. Crucially, however, I would argue that the line between "acceptable" and "unacceptable" autoposting does not just mean offering users the right level of control over content sharing or generation: it means taking seriously the autonomous and performative power of algorithmic personalization systems to bring selfhood into existence.

To return briefly to Jordan: even independently of algorithms, he argues that SNSs demand a new consideration of the ways that the self is invoked and constructed. Jordan states:

> Uneasily coincident [on SNSs] are the self as someone who comes to the network—in terms of private and public this is likely to be someone who comes with their identity as property—and the

> performances the identity puts on, which are required to exist on the network and so require publicness (2015, 130).

It is the "uneasy coincidence" of the self as preexisting the network, and the self brought into existence via networked identity articulation, that takes my interest here. This is because this duality of self is indicative of the ways in which algorithmic personalization demands that users be *both* endlessly expressive, produce "abundant" (Szulc, 2018) affective outputs across platforms, "express who you are through all the things you do" continuously via the commodifiable technologies of the self provided, *yet also* adhere to a "one identity model"—an anchored (Szulc, 2018), authentic, and "verifiable" (Hearn, 2017) self that can be affixed to an individualized profile. As explored in the previous chapter, the latter self is a model invoked by the Ghostery users interested in protecting "their scuzzy bits" (Chris, unemployed/activist, UK) from the dehumanizing threat of data trackers. In these Facebook user accounts, however, the self is more positioned as endlessly expressive, and entangled within the algorithm in ways that reconstitute identity.

The self that is performatively constituted through autoposting takes on a particular significance if we consider the cross-platform potentialities of autoposts to regulate and govern the identities beyond the boundaries of Facebook. If we take as an example Beth's regulation of her listening habits to suit the "publicness" demanded by Spotify/Facebook, it is possible to mark a tangible moment of constitution and regulation enforced by the personalization process—Beth's listening habits are redirected through a grammar of action that not only regulates her self-performance on Facebook but her performative articulations of selfhood as she listens to a (limited) playlist walking down the street. Furthermore, Beth's negotiations with Spotify reveal that apps can intervene in not only present articulations of the self but past ones, too—by adding an unwanted event to Beth's Facebook history, the Spotify app has the power to quite literally rewrite Beth's "past" selfhood on Facebook. Beth's account, along with the other participants testimonies featured here, highlights the performative power of Facebook apps to bring the self into existence, in ways that adhere to "legitimate" classifications of taste, public acceptability, and cultural interest, and which accommodate the commercial drive to anticipate and act on the user in the name of personalization.

6

Validating the Self through Google

Information that you need throughout your day, before you even ask.

—Google (2014)

They know I'm on campus, but I mean they got that wrong as well . . . I think like I'd trust them to be able to do more than that . . . I mean like with all the information and like algorithms and whatever they have, I wouldn't assume them to get it wrong.

—Lisa (student, UK)

This chapter is the last of three to explore how web users negotiate contemporary algorithmic personalization practices. The technology at the heart of this investigation is the "digital personal assistant" Google mobile app, formerly known as Google Now (Google, 2014). The Google digital assistant can be found pre-installed on Android devices across the world and is also integrated into Google's "smart speaker" device, Google Home (Google Home, 2018). Designed to respond to voice-activated commands as well as deliver personalized entertainment, scheduling, and product services, Google Home devices—along with competing smart speakers such as Alexa and Echo—have undergone an "explosive growth" in sales in recent years. In the last quarter of 2017, 18.6 million smart speakers were sold in the United States, the United Kingdom, Germany, France, and China (Watkins, 2018).

Making It Personal. Tanya Kant, Oxford University Press (2020). © Oxford University Press.
DOI: 10.1093/oso/9780190905088.001.0001

Though increasingly finding a literal home in domestic spaces as smart speakers, digital personal assistants were initially implemented on mobile phones—preceding hardware objects Echo, Home, and Apple HomePod are software apps such as Google Now. Though it has evolved into what is now simply "the Google app," this software should not be considered "dead": mobile digital assistants continue to be a key function of smart phones. It is Google's phone-based digital assistant that acts as the focus of this study.

Promising that "the more you use the Google app, the better it gets" (Google, 2018a), the Google app delivers real-time information deemed worthy of "*need* . . . throughout your day" (Google, 2014) in the form of "cards"—such as traffic and location updates, TV and movie recommendations, cinema times, "photo spots nearby," and stocks, sports, flights, and weather information. Unlike the established Google Search widely used in web browsers since 1997, the Google app initially promised frictionless, preemptive information retrieval that supposedly circumvents the need to actively input a search query in order to receive "relevant" information. Along with voice command recognition, digital assistants such as the Google app claim to offer algorithmically inferred suggestions that are automatically delivered to the user, and in so doing fulfill a user's needs without that user having to input a command or search query. Like all algorithmic personalization technologies, in order to achieve this, Google must collect and process user data deemed suggestive of a user's preferences, lifestyle, and identity. Though the exactitudes of what data Google harvests remain an industry secret, data collection broadly includes browser and search history (across devices) geo-location, purchase history, photographs, phone contacts, and email content—elements that Google tracks, collects, and stores about its users under the name of delivering them a "personalized" experience. It is through such systems of algorithmic inference that Google claims to be able to deliver to users "the information that you need throughout your day, *before you even ask*" (Google, 2014, my emphasis).

Google's predictive promise is at face value a neat rhetorical tagline designed to appeal to busy mobile users "on the go" (according to the study's participants at least, as I shall explore). However, the statement also exemplifies Google's long-standing conceptual ideal: to "know" a user's needs, desires, and intentions in order to provide the most efficient and convenient forms of search retrieval. In 2004 Google co-founder Sergey Brin envisioned that

> [r]ight now you go into your computer and type a phrase, but you can imagine that it could be easier in the future, that you can have just devices you talk into, or you can have computers that pay attention to what's going on around them and suggest useful information. (2004, cited in Levy, 2011, 67)

As established in Chapters 2 and 3, the idea that computational devices can preempt users' preferences, needs, and desires gives rise to potentially profound sociocultural implications for individual autonomy, knowledge production, and indeed a user's sense of self. Brin's sentiments, for instance, belie the fact that in order to "suggest useful information," algorithmic personalization systems must not only collect extensive user data but indeed *act* on and for users in their stead. In doing so, as scholars such as Hillis et al. note, information retrieval becomes akin to reading one's mind (2013), wherein the sovereignty of the user as decision-making actor comes to be undermined.

As I mentioned in Chapter 1, as mobile and home assistants have developed, producers have increasingly emphasized the voice-recognition capabilities that allow users to issue direct commands to their devices. During its initial development, however, the Google app was predominantly marketing on the strength of its "predictive powers" (Android, 2012): the app could retrieve information without the need to ask it to actually "do" anything. This emphasis signifies a shift in terms of the human/nonhuman agency model with which Google Search had previously operated, wherein the autonomy usually afforded to the user in searching for "relevant" information is instead afforded to the app—making the Google app an exemplary form of algorithmic personalization. This shift raises a range of critical questions relevant to this book. What does it mean to attempt to realize the task of giving users "what they want," "before they even ask"? What forms—and formats—of information does Google deem worthy of personal "need" to any given user's everyday trajectory? How does that user negotiate, appropriate, and understand this apparently necessary information? Indeed, in what ways, if any, is the user's lived experience informed, altered, and even constructed by Google's "predictive powers" (Android, 2012)? This chapter draws on the accounts of six Google mobile app users, who engaged with the Google mobile app over the space of six weeks, to explore these questions.

All participants in the "Google Now research project" (as the project was labeled during participant recruitment) were recruited from a Media Studies module at a UK university—all were recruited in the first term of their first year, and all were new to the university town and campus life. Their enrollment as digital media studies students raised a number of methodological considerations, the most pressing being that the content of one of their modules—which covered topics such as online privacy, search engine politics, and even the sociopolitical implications of personalization—was likely to inform and affect participant engagements with their app, and in turn their interview responses. To account for this, the study was conducted in the first few weeks of the students' first term at university.[1] As I will further explore, the mutual development of students' critical learning with their use of the app transpired to be a valuable site of exploration in terms of their engagement with the Google app. Participants' development of "expertise" intersected with their Google app use in nuanced, reflexive, and yet often contradictory ways.

These individuals were invited to participate on the grounds that they did *not* necessarily identify as "Google app users"—unlike the self-identified Ghostery or Facebook app users who participated in the previous two studies that constitute the empirical heart of this book. By asking these students to participate, this study hoped to capture the engagements of potential *non*-users of an algorithmic personalization technology. As Baumer et al. note, the "non-user"—that is, "a particular individual or group of individuals . . . [who] are unable or choose not to use some specific technology or technological system" (2015, n.p.)—is perhaps the most elusive mode of socio-technical subjectivity to identify and theorize. Their very status as individuals who *don't* use, or do not consciously identify as using, a particular technology often mean that non-users remain "in the margins" or indeed somewhat invisible as subjects of research. Yet Baumer et al. find the study of non-use valuable exactly for this reason, because "focusing explicitly on non-use can function as a dialectic, an inversion that provides a novel perspective on, and potentially fuller understanding of, the complex, multifaceted relations among society and technology" (2015, n.p.). Though two participants had limited interaction with the app prior to the study, and all participants did end up "using" the Google app as part of the study, I approach their engagement through the lens of "non-use" because participants could not find a function in their everyday lives for the app, as I will explore.

In order to explore the idea of "finding a use," it was important to gauge the level of engagement that participants already had with the Google app prior to the study. Two of the students—Tariq and Giovanni—had previous, very limited engagement with the app, though both told me that they did not use it often. The other fours students—Laura, Rachel, Lisa, and Heena—had never used the app.[2] These four students were split in two sections: two participants, Heena and Rachel, who could customize the app in whichever way they wished and confirm the app's inferences, and two participants, Laura and Lisa, who were at first not allowed to customize the app or confirm any of the app's predictive interferences. Acknowledging and managing the level of usage that these participants had with the app transpired to be a crucial component in some participant responses—for example, in Laura's insistence that the app would work "better" if she had been allowed to customize it, as explored later. Once participant use was established, I asked each participant to check the app at least twice a day and send me screenshots of any of the "cards" the app delivered to them. We then met four times over the space of six weeks to discuss the app's inferences and predictions. Four of the participants were interviewed together as a group (Tariq, Lisa, Rachel, and Heena), and two participants were interviewed separately (Giovanni and Laura, each in one-to-one sessions).[3] The following sections explore some of the themes, negotiations, and responses that emerged from interview sessions. They are broadly discussed in the chronological order.

"COOL," "IMPRESSIVE," "SMART," BUT "USEFUL"? FINDING FUNCTION IN PREDICTION

To turn to the study itself, for all participants Session One (and each subsequent session) began with the question *"What cards has Google Now shown you this week?"* In the first session, this question elicited the general consensus that the app mostly showed the participants local weather updates, traffic information to and from "home" and "work," and "places nearby" (these "places" being exclusively restaurants and cafes). The most frequent card shown to participants was local weather, followed by a "home" to "work" commute time, which displayed a mode of transport (car, train, bus) inferred by Google based on the GPS location and travel trajectory of the users in question (see Figures 6.1–6.3). In this first session, participants' overall responses indicated a sense of disappointment: all participants said that they thought the information delivered by the apps' cards was lacking or "not enough." For instance, Laura stated that Google had shown her weather, restaurants, and "my location . . . but that was about it," while Lisa also stated, "Yeah I just got the weather." For participant Rachel, the app proved similarly disappointing: in both Sessions One and Two she stated that the app showed her "absolutely nothing" (apart from the weather)—a source of much frustration for Rachel, as I will explore shortly. Tariq also echoed the sentiment that the app did not offer much, stating, "it just gave me information about the traffic."

In Session One, then, the app did not seem to be offering the participants much in terms of information that they "needed throughout their day"; traffic, location, and weather updates did not in these initial stages seem to fulfill their desired use or expectations of the app. However, this failure to deliver "useful" information did not stop many of the participants from

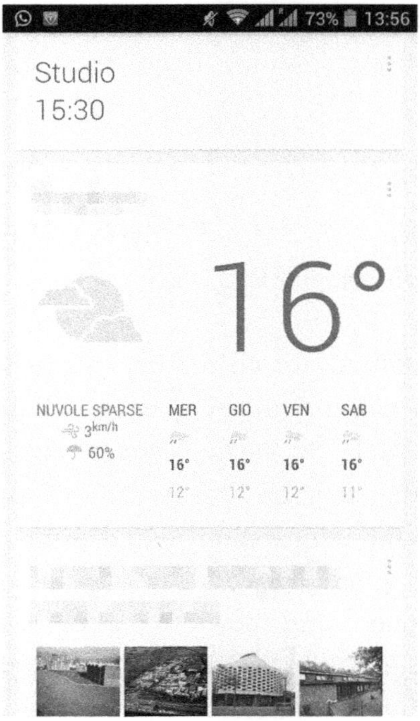

Figure 6.1. Giovanni's screenshot of the local weather.
Credit: Screenshot by Anonymous Research Participant (pseudonym Giovanni).

expressing positive sentiments. When I asked Laura if she had found the "nearby places" card to be helpful (which showed her Japanese restaurants near to her location), she replied:

> LAURA. Yeah it was helpful. It was helpful to have it there.
> *Interviewer: Did you use any of the restaurants?*
> LAURA: No [laughs].

Laura's statement that the Google app was "helpful" was quickly contradicted by her self-conscious answer that no, she did not in fact actually use the information that was provided to her—she went to an alternative restaurant instead. Similarly, Heena expressed positive sentiments regarding Google's predictive powers:

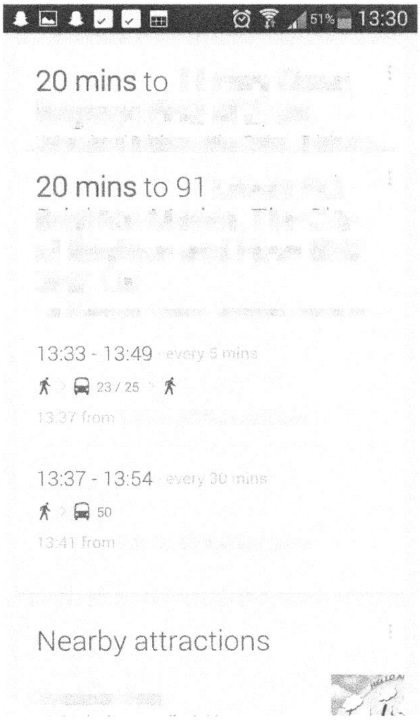

Figure 6.2. Tariq's (incorrectly) inferred "commute" from "home" to "work" (he both lives and "works" on campus).
Credit: Screenshot by Anonymous Research Participant (pseudonym Tariq).

Interviewer: How do you think [the app] manages to predict where you work and where you live?

HEENA: I think it predicts quite well because I mentioned that I was in Southville[4] but then it managed to guess that I was in Southville Lane which is quite good, and then it guessed where I had my classes, which is more interesting.

Aside from the fact that Heena does not really address the question (instead of explaining *how* she thinks the apps predictions work, she states that Google predicts "quite *well*"), Heena seems to be most interested by the app's inferences, which she deemed to be "quite good." What is determined to be "good" here is simply the fact that Google has predicted

Figure 6.3. Laura's "local weather" and "places nearby" cards.
Credit: Screenshot by Anonymous Research Participant (pseudonym Laura).

something correctly—it is not that the app has been "helpful" by showing Heena where she has to go. After all, Heena already knows where she is, and where her classes are: Heena's praise seems to lie in the fact that Google knows this, too.

These positive sentiments were also expressed in Session Two. For example, Tariq, who in Session One had also found that Google mostly just showed the weather and locations nearby, was excited to discover that this week the app had correctly inferred he had taken the train the day before:

> TARIQ: Oh look! It's got the train station on there now because I took the train yesterday! That's good!
> *Interviewer: Are you taking the train today?*

TARIQ: Um, no I don't think so, I'm not going to, well I am going to go into town but I think I'm going to take the bus.

The fact that Google's cards display information that is ultimately not useful to Tariq seems to be inconsequential—he remains impressed by the app's offerings, despite this lack of tangible use.

Here, then, it is the app's predictions *in and of themselves* that are deemed by Laura, Tariq, and Heena to be "good," "smart," and "impressive." It seems that these participants' positive sentiments refer to Google's dataveillance strategies and subsequent predictive *powers*, rather than the helpful *outcomes* or tangible usefulness of those predictive powers on their everyday trajectories. It is the possibilities for anticipation that the app offers, rather than the realization of those possibilities, that seem to trigger positive comments.

How can these positive sentiments—even in the face of disappointment—be theorized? Broadly speaking, as Wyatt notes, emerging technologies are often publicly framed by institutions, media, policymakers, and even as users as "necessarily desirable" (2005, 68) in ways that set up the technological objects as implicitly functional: in simply existing as a new object, technological devices are assumed to be useful. Similarly, Best and Tozer's qualitative study on user resistance and appropriation of new technologies identifies a process in which users find a way to accommodate new socio-technical into their everyday lives, even as such objects fail to be useful (2012, 14). Thus it is the predictive possibilities of the app that are demarked as "cool" and "impressive," even as the app fails to function in a way that these participants expect. I do not mean to suggest, however, that if the app *did* function as "useful" then a critical interrogation of the app would not be valuable. As explored in Chapter 2, critical interrogations of the "successes" of algorithmic personalization expose a host of sociocultural implications in regard to user control, commodification, autonomy, and knowledge production—even when personalization fulfills its expected functions to preempt the users' needs as desired. I will explore the critical implications of the Google app's "successes" later in the chapter, but here it seems that these participant responses reflect a wider discursive tendency to frame new technologies as implicitly beneficial, even when the technology fails to offer up functions that meet the expectations of the user.

SELF-BLAME: THE TRUST THAT GOOGLE WILL PROVIDE

As well as finding positive sentiments in Google's ultimately unhelpful inferences, during the first few weeks of the study participants offered a range of explanations as to why the app's interface lacked useful information. For example, in Session One, Tariq attributed Google's incorrect prediction of his "home" address to poor mobile internet signal, and Rachel stipulated that her inability to connect to the campus Wi-Fi might be the reason that Google Now was showing her "just the weather." Alternatively, Heena offered the following reasoning:

> Well, for some reason I only have the weather, but I guess it's because I'm a new user, and I don't use it a lot and I don't even customize it.

According to Heena, the app's apparent failure to provide useful information was not due to a failure on Google's part; it was rather *her* status as a "new user" that was responsible for any technological shortcomings. Lisa also blamed her lack of engagement with the app for the app's failure to deliver more cards, stating that "I'm not using mine properly, if I had more in it then it would be helpful," referring here to the fact that, as part of the study's initial conditions, she was not allowed to customize the app in any way. Similarly, in Session One of her one-to-one interview, Laura seemed adamant that her issues with Google could be put down to the fact that she was not allowed to personalize the app, as evident during the following exchange:

> *Interviewer: Why do you think that Google Now chooses to show you things like where you work?*

LAURA: Probably for the convenience factor, and to make it more efficient I guess, to show you different routes to get there and to show you personalized ways to try and help you get to places. . . .

Interviewer: Has it actually helped you find anywhere?

LAURA: No, no [laughs] it said where I worked but it didn't show me how to get there. Yeah I think that's because I didn't put any of my personalized details in, I think when I do it probably will.

By Session Two, Google had not shown her anything more than weather and travel updates. Again, when I asked her if she used the app that week, she replied:

> Just, actually just for the weather because there wasn't anything that came up at all, apart from yeah, that was really it. But I think I would've used it a lot more if I did have, if I was able to put in stuff that was relevant.

Due to the app's lack of cards and Laura's assertions that the study was limiting her use rather than the app, we agreed that at the end of Session Two, Laura could customize the app in any way she wished in order to try to improve the app's functionality. Consequently, Laura's hopes were in some ways fulfilled—her user-initiated customization of the app resulted in the display of many more cards. For example, in Laura's third interview, she told me that during a visit to the airport for a trip abroad Google had accessed her email and subsequently showed her a "card" with her flight details—"so it's basically saying that my flight was on time and stuff, and all the details about the flight, and what terminal I'd arrive in and what time, so I found that quite cool, I was quite excited about that." Again, however, the idea that Google's predictive powers are "cool" in and of themselves still persists—she stated that the app "was really efficient, I was in the airport and I opened it and it was just out of curiosity and it said like all the things that I needed about my flight and it was really interesting," suggesting that though the app's inferences were "interesting," she did not actually utilize them in catching her flight.

The reasons offered by participants for the app's initial failures seem thus to displace the "blame" for Google Now's lack of functionality anywhere *but* Google itself—the app has failed to provide because of 3G; it has failed because of the campus's Wi-Fi network; it has failed because of the disengagement of these users *themselves* with the app. In his analysis of tablet devices in labs, Burns describes a similar displacement of blame invoked by technological adopters in which, in order to retain their perception of tablets as "perfect" techno-social objects, "users . . . responded to unexpected failures of their devices by tactically redrawing the boundaries of the object so as to eject the faulty element" (2015, 9). A similar process is at work in participants' articulations regarding the Google app: the "boundaries" of the app are consistently redrawn in participant explanations as to why the app is failing, in ways that eject any disfunctionality to outside of the app itself. In the context of this study, participants' redrawing of the Google app is contingent on the epistemic uncertainties that I argue are inherent in contemporary algorithmic personalization practices: users do not know exactly how personalization functions—or indeed *does not* function—and so the "faulty elements" of Google's personalization systems can be displaced to their mobile networks, or Wi-Fi, or indeed the users themselves. This redrawing to eject the app's faults did not last, however; as I will explore, Google was eventually taken to be (still somewhat mysteriously) at fault.

Burn's analysis is valuable in understanding broadly how technological objects might be bounded by users as "perfect." For the purpose of this study, however, it is worth interrogating the role that Google plays as a specific socio-technical force. Since its inception, the Google search engine has enjoyed widespread global uptake: as of July 2018, Google Search took a 71% share of the search engine market on computers and an 88% share on mobile devices, meaning that the majority of web users use Google as their primary search engine (NetMarketShare, 2018). Combined with Alphabet's (Google's parent company) extensive range of Google-affiliated products, Google has an extraordinarily dominant place in the digital economy. However, Skeggs makes an interesting distinction regarding Google's place in global markets: Alphabet's venue generation comes largely from "non-productive areas of the economy": that is, from the "circulation of revenue" (2016) through advertising and information consumption, rather than the production of material goods (as supposed to say, Apple, which dominates through material goods such as the iPhone). The non-material dominance of Google is not just economic, however; it can also be considered at a symbolic level, for example by the now well-reported fact that in 2006 "Google" was added as a verb to the Oxford English Dictionary. The fact that web users no longer "search" for information—they "Google" it—is significantly suggestive of Google's naturalized place in the digital everyday; it is *the* search engine, its worldwide use so ubiquitous as to have become ideologically concretized as a socio-technical practice.

Google's increasing naturalization works to reinforce the company's power as a socio-technical force, as Noble (2018), Jarrett (2014), and Hillis et al. (2013) have argued. For instance, Noble (2018) notes that users worldwide trust Google to provide "objective" and "neutral" search results in ways that mask the commercial biases inherent in its search engine protocols, as I revisit later in the chapter. However, here I want to highlight that the dominance of Google as *the* search engine seems to be invoked in participant responses. Their trust in Google lies on the assumption that when it comes to information retrieval, Google should—and does—just *work*. This is reinforced by participants' own testimonies regarding search engine use: though most had never used the Google app before, all participants reported that they used Google Search (on both their computers and mobiles, through browsers such as Chrome) as their only search engine.

Of course, the fact that Google Search has for the past few decades successfully met its users' expectations does not exempt the company from critique—Hillis et al. highlight that the very ideas of "efficiency" and "convenience" are themselves "meta-ideologies of the contemporary technicized, consumerist conjecture" (2013, 5). In participant accounts so far, Google's "efficiency" and "convenience" are rendered questionable by the lack of information the app has provided to participants; and yet participants continue to trust that Google *can and should be* capable of providing an efficient and convenient service. It seems, then, that Google's reputation not just precedes itself; it actively constructs and manages these user expectations, even in the face of failure.

PRIVACY: THE TRUST THAT GOOGLE WILL PROTECT

In the first two sessions, the apps' front-end capabilities were described as "impressive" yet consistently failed to live up to the participants' high expectations. I will return to these expectations shortly, but first I would like explore some of the back-end data tracking capabilities of the app. For all participants, the fact that Google *was* tracking them in some way was established very early in the study: all participants knew that Google could track their geolocations and search history. How then did they feel about their positions as "data providers" (Van Dijck, 2009)?

It quickly transpired from Session One onward that the app showed more cards to Heena than it did to any of the other participants—even though Heena had not customized the app or confirmed the correctness of any of the app's inferences. Heena was perplexed that Google seemed to "know" a lot about her—for example, Google seemed to be capable of tracking her online television watching and inferring shows based on her viewing habits (see Figures 6.4 and 6.5). As well as calling her (comparative) bounty of cards "smart" and "interesting," Heena also described the app as "like your own personal stalker." When asked *"Why do you feel like that?"* she stated:

> Because it can tell you where you're going and because it can assume what you're like and I think it's kind of scary at times because . . . maybe it can use your information like against you it's like identity theft and it's like really scary, and identity is all you have, it's yours, and it's like a little bit scary.

Heena's description of the app as "scary" was offered in Session One alongside more positive descriptions of the app's predictive powers. This

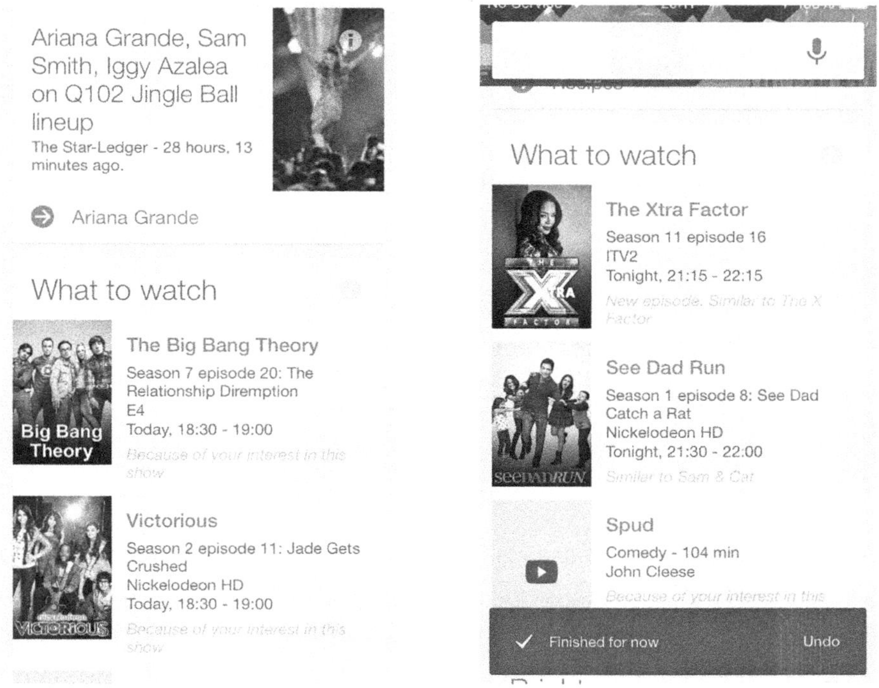

Figures 6.4 and 6.5. Heena's "what to watch" recommendations.
Credit: Screenshot by Anonymous Research Participant (pseudonym Heena).

somewhat contradictory juxtaposition persisted throughout the study—by Session Three, Heena was still describing the app as a "personal stalker," and she also expanded on what she meant by "identity theft." For instance, she stated that she would not want Google to know her permanent address:

> Who knows, maybe there's someone in Google, like hackers, like they might just be able to track someone down, like they know everything about you, so they might know like incriminating things, and then they might want to target you. So if they have your permanent address, then you're screwed.

Here Heena seemed to be less concerned by Google's own treatment of her personal data, and more worried that "incriminating things" about her might fall into the hands of "hackers"—Google *itself* is again framed as trustworthy. Similarly, in Session Three, Tariq stated:

> Literally what would happen if I gave up all of my information to Google, I feel like I have nothing to hide, but then there's certain things that I do that I wouldn't want my parents to know.

In some ways, Heena and Tariq's sentiments echo the findings of wider studies that analyze young people's responses to online privacy invasion (Lapenta and Jørgensen, 2015). As Lapenta and Jørgensen note, the notion that young people are "digital natives"—that is, users who are "born" into digital media technologies and therefore equipped with the expertize needed to successfully navigate them—is often mobilized in policymaking and media accounts to suggest that for young people, privacy is "no longer the social norm" (Lapenta and Jørgensen, 2015, 4). They note, however, that there is a building body of research suggesting this is not the case: "On the contrary, these studies of young people's practices on social media platforms illustrate the emergence of a new privacy norm that corresponds to the structural conditions of online social life" (2015, 3). These emerging "structural conditions" tend to frame privacy as a matter of control over their *self-representation* online—young users predominantly seek control of their own photos, identity expressions such as status updates, and indeed representations of their selfhood more generally. As such, Lapenta and Jørgensen argue that privacy risks for young web users tend to coalesce around user-to-user hacking or identity theft, "whereas potential privacy risks related to the state or private companies receive limited attention" (2015, 7). This is at work in Tariq and Heena's responses: despite Google's extensive data-mining strategies, it is not Google itself that represents a threat. Instead, it is threats to their self-representation that are framed as more pressing: either from hackers looking to steal Heena's identity, or from Tariq's parents, who might look to discipline or condemn their son's self-disclosures. Google as a commercial, socio-technical force is discounted from any privacy concerns—Google will protect, even though it is responsible for the data-tracking practices that, if "misused," pose a threat.

THE DATA-FOR-SERVICES EXCHANGE: THE TRUST THAT GOOGLE IS WORTH IT

Though commercial privacy concerns were not perceived as substantial, throughout the study participants acknowledged Google's mobilization of the data-for-services exchange that, as established in earlier chapters, is commonly conducted between users as data providers and platforms as service providers. For example, in Session One, I asked Tariq if he minded that Google showed him pictures of "places nearby," and his response almost immediately became a matter of data submission in exchange for the app's services:

> TARIQ: For me it's not really a simple answer, so for me so like it's really useful and handy sometimes, like . . to tell you what time at home it is because I remember one time I called my parents on Skype it was like 9 here but I realized it would be 12 there because of this, so they probably wouldn't pick up, but um it's useful in that way but then you're also sort [of] surrendering a part of yourself up to Google, right? So, they've got all this power over you, they know so much about you, so it's just really complicated I think.
>
> *Interviewer: When you say surrendering, what do you mean by that?*
>
> TARIQ: Like basically it's information about yourself that Google could potentially use to take advantage of you, I guess that's what I mean by surrendering—well you see I'm just like one person in 7 billion so what interest would they have in me which is why I mostly don't care.
>
> *Interviewer: Do you ever feel taken advantage of?*
>
> TARIQ: By Google? No.

Here Tariq feels that he must weigh the convenience and benefit of using the app against the threat of "surrendering yourself to Google." He rearticulated in Session Three that allowing Google to track him involved a negotiation between feelings of "apathy" and "fear" of Google's data-tracking practices—an uncomfortable juxtaposition also articulated by Laura, Lisa, and Rachel. For example, Laura, who throughout the study was extremely positive about the app, told me in Session Four that she felt that Google targeted "younger people," who could then be left vulnerable to "manipulation" by Google. She stated during the following exchange:

LAURA: . . . we're more reliant on [Google] I suppose and we use it for a lot more things than older people would I would say, and so we're more, we don't have as much knowledge on it so therefore we kind of don't really realize how it's manipulating us.

Interviewer: Do you mind about that sort of thing?

LAURA: No, not really.

Interviewer: Why don't you mind?

LAURA: That it manipulates us?

Interviewer: Yeah.

LAURA: Well, I think it's mainly that they manipulate us in order to, like for marketing purposes I suppose, and for advertising, to personalize it, so it's not really to invade your privacy on purpose, it's for their own business purposes in order to benefit the company, it's not to stalk you, so [pauses] and anyway it's helpful.

Of all the participants, Laura displayed the strongest faith in the app and stated that she "definitely" trusts Google. Laura's, Heena's, and Tariq's sentiments support another point made by Lapenta and Jørgenson in regard to young people's treatment of privacy online. They state that for young people "the 'repurposing' of their data (data mining, commercial use) is perceived as a precondition for social participation" (2015, 8). In other words, "signing off privacy rights to the social network was seen as a necessary price to be paid in order to participate" (2015, 8). However, I want to emphasize that Tariq's and Laura's responses clearly draw parallels with the data-for-services negotiations that some of the Ghostery interviewees also recounted—despite self-identifying as "privacy concerned" web users, the Ghostery users interviewed were, like these Google app participants, acutely aware that their data were harvested and repurposed as a pay-off for free-to-use services. This suggests that not only "young people" partake in these exchanges: it is *all* web users who find themselves in this trade-off, even those resolutely interested in protecting their privacy.

A FAILED EXCHANGE: AN EXPERIMENT IN DATA-FOR-SERVICES

How does this exchange of data-for-services intersect with the participant accounts of the app's failures? Though this sacrifice of privacy for Google's services was deemed "worth it" for most participants, there is a disconnect here between participants' acceptance of the data-for-services "contract" with Google and their assertions that the app was not providing the level of service they expected: the app was consistently showing little or nothing that they found helpful. When it came to providing a personalized service in exchange for personal data, Google did not seem to be holding up its side of the bargain, so to speak. To explore this, I want to briefly expand on Tariq's engagement with a "card" that he *did* find useful: the app's "currency converter" card (see Figure 6.6). Tariq, who already had Google activated on his phone prior to the study, stated that he finds this card particularly useful:

> Right now [the app] still thinks I'm in Dubai, Dubai is where I live, so it thinks I live in Dubai but it's says like, it shows me tourist hotspots, and there's like this really handy currency converter.

It became apparent that showing the currency converter was an attribute specific to Tariq's usage of the app compared to other participants' usage—and as he states here, Google showed him the currency converter because Tariq's "home place" was set in Dubai—Tariq's home city. Thus, Google

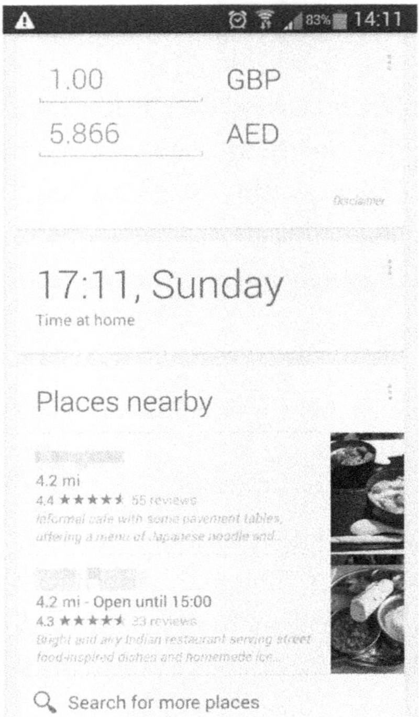

Figure 6.6. Tariq's currency converter card, displayed at the top of the screen. Credit: Screenshot by Anonymous Research Participant (pseudonym Tariq).

assumed that Tariq was a tourist visiting the United Kingdom, and therefore in need of a currency converter.

The currency converter was indeed useful to Tariq, and yet it seemed that the only reason Google had shown him this card was *because* it had wrongly inferred that Tariq was a tourist, rather than an overseas student. Tariq and I speculated that Tariq's currency converter might disappear if Google correctly inferred that he now lived in the United Kingdom, which up until Session Three it had not inferred. Given the lack of engagement that the app had so far afforded to the participants, it seemed fruitful to try to make the most of this useful card; that is, to perform an informal experiment to see if the currency converter card would still appear if Tariq's "home address" were set to his university address. During Session Three it was decided that we should address this hypothesis:

Interviewer: Did you end up resetting your home address?

TARIQ: No I was still kinda hoping that [Google] would but I'll do it now.

Interviewer: Is that all right? Can we see, can we see what happens if you reset it?

TARIQ: Sure.

[Tariq manually sets his "home address" to his campus address]

Interviewer: Do you still have the currency converter?

TARIQ: Um no, oh well, I can always find it I guess.

. . .

Interviewer: How do you feel about the fact that that information is not there any more?

TARIQ: Um, well hopefully, it will give me other helpful information. But I can always like customize it so I have the right cards, right?

Again displaying faith in Google's "predictive powers," Tariq had been hoping that the app would automatically infer his new "home address" to be campus; but as the app had not yet performed this, Tariq was happy to initiate it and input his campus address. However, once Tariq had input the "right" and supposedly relevant personal data into the app—that is, his "correct" home place—the information that he had found most useful subsequently *disappeared*. In being listed with a UK home address, the app had assumed a tourist-oriented card was no longer "needed" by the user.

Despite the card's disappearance, Tariq remained optimistic that the app "will give me other helpful information"—and hoped that he can "always find" the currency converter by customizing the app to display it once again. However, after some investigation, it transpired that he could not find a way to customize the app and display the currency converter. Ironically, then, Google's attempts to provide "personally relevant" information through individual inference actually had the opposite of the intended effect: guided by the (in this case flawed) logic of algorithmic personalization, the only "card" that Tariq found useful to his lived experience was stripped from the app. In this instance, Google's role in the exchange of personal data for personalized services is made highly problematic—the "personal" information that Tariq gave up to Google actually resulted in a less useful, less convenient, and less personal service.

"I'VE GOT SO MANY INTERESTS!": THE TRUST THAT GOOGLE "KNOWS" YOU

By Session Three, it was firmly established that Google did track participants. However, the specifics of exactly what the app "knew" about them remained uncertain. As part of Session Three, I therefore wanted to discuss with participants the extent of their knowledge surrounding Google's attempts to track them. Despite the pledges of transparency made toward web users in accessing their own data trails, it is notoriously difficult to gain comprehensive data "back" from contemporary data holders (Brunton and Nissenbaum, 2015; Pasquale, 2015). Google is no exception, although one tool that they do offer in regard to user data is Google Ad Settings. Google Ad Settings displays to Google users their inferred gender, age, language, "topics that you like," and "topics that you don't like" (Google Ad Settings, 2015), with these categories mined and largely algorithmically determined from Google users' search history, YouTube history, and Google+ profile—though again, the specifics of how a user's Google Ad Profile is aggregated are not publicly available. Though only able to offer an epistemological glimpse into Google's complex and abstract dataveillance strategies, Google Ad Settings enabled a conversation between myself and participants regarding the back-end data-collection strategies deployed by the company more broadly, if not the Google app itself.

Initial access to Google's Ad Settings elicited mixed reactions from participants. Giovanni was concerned that Google had inferred interests for him, while Laura was happy to be profiled in such as manner, as long as it led to "conveniently" personalized results. Conversely, Rachel's reaction upon seeing her profile was at first one of profound disappointment, as all of her "settings" were blank: Google did not know her age, gender, or interests. As stated earlier, Rachel had routinely expressed frustration that the app had failed to show what she felt were useful cards over the duration of the study. As she explained in Session Three, the lack of cards occurred despite her best efforts to integrate her preexisting networked services into the app:

> [The app is] linked to my YouTube, like I checked what it's linked to, and I like checked my privacy settings, and so I go on YouTube and like nothing pops up and like I literally ask it questions . . . but it won't give me cards.

The disappointment Rachel felt toward the app's functionality was also reflected in her interactions with the Google Ad Settings. However, this time, her despondency had little to do with Google Now's lack of *function*; instead, Rachel was disappointed by the fact that Google apparently "knew nothing" about her—her settings were blank.

Rachel's efforts to connect to other services in order to be "noticed" by Google draws parallels with some of the data-oriented "tactics" used by other participants featured in this book. In attempting to be recognized by Google, Rachel is employing a tactical maneuver in that she tries to turn to face the algorithm (Gillespie, 2014, 184) to suit the computational protocols of the app. However, I use "tactic" cautiously here. After all, Rachel is clearly not attempting to evade, challenge, or resist Google's algorithmic personalization mechanisms in the De Certeauian sense established in Chapter 4, where Ghostery users knowingly employed playful, emotive, and potentially futile "tactics" as a form of resistance against the oppressive, omnipresent surveillance "strategies" (De Certeau, 1984, 39) deployed by data trackers. Instead, Rachel's efforts to connect to the app render her a "tactician" who is attempting to turn *toward* algorithmic power. She cannot know for certain what maneuvers she must make to suit the operational logic of the app, and so she must, though "lacking a view to the whole" (De Certeau, 1984, 39) of Google's algorithmic operations, try to connect her interactions on other platforms to ensure that she is noticed by Google.

So far, over the study, Rachel's attempts have not succeeded—as she states in the preceding, she simply could not get the app to show her cards. During the course of the Ad Settings exercise, however, Rachel's "luck" changed: she realized that she had omitted to sign into her Google account, and that this was affecting what she could see. On signing into her Google account, her Ad Settings profile was suddenly revealed:

RACHEL: Oh, it does know I'm female! That's nice!
[She finds here inferred interests]
RACHEL: Oh, and I have got interests! I've got so many interests! . . . I've got so many good ones!
Interviewer: What good ones have you got?
RACHEL: Um, like ones that are just actually like, um, are like me, I've got like loads of animal ones, like dogs, wildlife, which I'm super into. I've got rock music, fashion and style, hair care, I've got oh, I've got make-up and cosmetics. I've got metals and mining, and I don't ever do that . . . I've got like five out of 65 that I'd don't do, but the rest of them are pretty good.

Rachel is obviously pleased about the accuracy of Google's algorithmic inferences. More than this, though, Rachel seems happy that this specter of her "algorithmic identity" (Cheney-Lippold, 2017) simply exists:

Interviewer: So how do you feel about them?
RACHEL: I'm so happy [laughs] that I've finally got something from them, like actually.
Interviewer: How come it makes you happy?
RACHEL: I just feel really excited, I don't actually know [laughs] I think it's, for ages it didn't like do anything, and I was just really disappointed, like I expected a lot from it so I'm quite happy now, at least it knows my interests.

It is important here to consider exactly what Rachel feels she has "got from them"—"them" referring her to both Google Ad Settings *and* the Google app. Rachel has in fact received nothing more from the app in terms of correct functionality or service —all her "Ad Settings" profile has revealed is that she has been "noticed" and profiled by Google's wider profiling mechanisms. Rachel's pleasure at finding her Ad Settings seems to negate the previous disappointment at the Google mobile app's lack of functionality—yet, as is clear from her previous disappointment, her Ad Settings profile has apparently *not* been used by Google to inform or enrich her experience of the Google app. As Rachel herself states, "it's weird because like I didn't know like it [Google Ad Settings] knew like so many of my interests, so it's like it's weird that it [the Google app] doesn't act on that, like it doesn't actually show me cards related to that sort of stuff." There is a clear disconnect here between Google's tracking of Rachel's intentions and the personalized data outputs that Rachel is (supposed to be) receiving in the form of the Google app's "cards." Yet Rachel remains happy that, at the very least, Google as a broader social-technical entity "knows" who she is.

Rachel's pleasure in being identified by Google can be theorized as a moment in which techno-social objects facilitate self-recognition. Focusing on mobile phones as such facilitators, Mowlabocus argues that due to their intimate connection with individuals, mobile devices often act as "transitional objects . . . [that] serve as negotiating points between self and other" (2014, 1). Mowlabocus uses Lasch's work on the culture of narcissism to explore how mobile technologies might "secure our sense of self" (2014, 1) in twenty-first-century individualized frameworks. He states:

> In our phones we become consumed by recognition of ourselves through the eyes of others (tagged comments, liked posts, other forms of phatic communication) that become mediated, brought to our attention by these technologies. (2014, 2)

Mowlabocus refers largely to a process wherein the mediated interactions in which human actors engage constitute moments of recognition. However, unlike the peer-to-peer forms of validation described by Mowlabocus in the preceding, Rachel's pleasure in self-recognition comes not from the eyes of others *but from the presentation of her "algorithmic identity" as constituted by Google.*[5] The self is secured and rendered readable through wholly algorithmic mechanisms, which here have been configured through "gender, age, and interests."

Despite the presentation of Google Ad Settings as a "personal" profile, as established in Chapter 2, the algorithmic identity presented to Rachel has little to do with her being explicitly "known" as an individual by Google. As Andrejevic (2013) and Cheney-Lippold (2017) stress, commercial data aggregation is not interested in *individuality* but in dividualized patterns of mass user behavior that essentially "disassemble" the subject in order to computationally grammatize (Agre, 1994) and commodify the self through algorithmically recognizable and value-generating data sets. From Rachel's perspective, however, her Ad Settings reflect her individual interests and her personal identity markers—an entirely understandable interpretation given that her Ad Settings represent Rachel as a holistic, individualized subject who has a defined set of interests.

As Bolin and Andersson Schwarz's (2015) work suggests, and as I detailed in Chapter 2, the individuality constructed in Rachel's Ad profile instead is (re)constructed from dividualized data aggregation: abstract, correlational data points are "*translated back*" (Bolin and Andersson Schwarz's, 2015, 1, my emphasis) from Rachel's data trail into categories that Rachel herself can recognize as applicable to her. In the translating back of dividual data sets into a figure of the individual, Google Ad Settings facilitates a moment of self-recognition where the self is momentarily and algorithmically validated, legitimized, and fixed into a personal profile that Rachel recognizes as not just "accurate" but pleasing.

Rachel's engagement with her "algorithmic self" indicates the apparently "truthful" but also epistemologically mystical forms of knowledge that algorithms produce in everyday life. As Finn notes, part of the "magic" of algorithms is that they produce new forms of personal intimacy: he argues, "we desire that algorithms truly know us and tell our stories" (2017, 75). Though Finn applies this idea more broadly to algorithms in popular culture, this sentiment is particularly resonant in personalization algorithms that quite literally seek to identify users in order to "know" them in and through data, in order to "tell our stories." These algorithmic stories are told to two sets of listeners: firstly to marketers and businesses interested in monetizable algorithmic profiling mechanisms, and secondly to users themselves in the form of profiles such as Google's Ad Settings. Such forms of algorithmic validation thus work to legitimize the web user as a "doing subject" from the users' perspective, even as they obscure the dividuating and dehumanizing processes inherent in contemporary algorithmic personalization practices.

EPISTEMIC TRUST: THE FAITH THAT GOOGLE CAN PERSONALIZE

As the study continued, participants' expectations of the app's capabilities remained very high, even as the app failed to meet those expectations. Similarly, on a number of occasions participants seemed to expect the app to "know" and "anticipate" their actions—and therefore algorithmically personalize the app—to an extraordinarily high degree. For instance, during Session One, Heena pointed out that so far, she had not used the "weather" card because she had other means of telling the weather at her disposal:

> HEENA: I don't check the weather, I just open my window and just feel how it feels outside.
>
> [Everyone laughs]
>
> *Interviewer: Why do you think Google thinks that's helpful [to show the weather]?*
>
> HEENA: I don't know really, maybe it thinks I'm in a foreign country, because I did come from Malaysia where it's generally very hot all the time, so I think I come here and it just wants me to know that it's going to be colder in England, and for me to put a few more layers on [laughs].

This light-hearted observation omits that fact that the app shows *all* participants the weather (as already established, even in Session One of the study)—however, Heena assumes that the weather card must somehow have identified her, knows she is from Malaysia, and therefore anticipates that she will be cold in England. In the same session, the "work" feature of the app was subject to similar speculation. The group discussed why Google had inferred "work" places for the participants, even though the participants are all students and therefore perceived themselves to not be engaged in the act of "work."[6] Tariq and Lisa deliberated why Tariq was allocated a "work" place by Google (bearing in mind that Tariq used the app prior to the study):

TARIQ: I do remember that even back home when I used to go to school . . . [Google Now] didn't say school it said work, time from home to work, distance from home to work . . . so it thought you were an adult.

Interviewer: Yeah, so you think it should say school?

TARIQ: Yeah, but it never really asked for my age did it, I have a Google account, but I'm not sure if it's got my age on it.

LISA: I think you might do because they put Google and YouTube together at some point, and YouTube generally wants your age in case there's like a video that's 18 or over, so like maybe they got it from work.

TARIQ: Yeah, yeah, yeah, because back in the day, well a few years ago, when I was 18, I pretended I was 18 when I joined YouTube, so that's why it thinks I'm old enough to work or something.

Tariq assumes that the app infers that he goes to work because he once pretended he was 18 years old in order to join YouTube; and Lisa proposes that Google has obtained this information via Google's 2006 acquisition of YouTube. To give one more example of such speculations, in Session Two, Heena told the group that Google Now had shown her the "stocks" card that week, which led to the following exchange:

Interviewer to Heena: Why do you think Google Now shows stocks?

HEENA: I don't know.

TARIQ: Is your dad into investments, or your mum, like?
HEENA: Yeah, probably, but how would [Google Now] know my dad was?
TARIQ: Oh wait, he's not on Facebook? [As established earlier in the focus group]. Google plus maybe?
HEENA: My dad doesn't have an account.
TARIQ: Oh, OK.

Here Tariq speculates that Google might know that Heena's father is "into investments" *and* has access to Heena's Dad's Facebook or Google plus account *and* might know that Heena and her Dad are related, and therefore is be able to infer that Heena is interested in stocks. The speculation is dismissed by the conclusion that Heena's dad does not have Facebook or a Google account, but no other explanation is thereafter offered by any participant.

There are two points I want to make here. First, in speculating about the app's predictive and personalizing capabilities in this manner, these participants (like Rachel) display an epistemic uncertainty about what data the app is mining from them across platforms, when, and how. They know that Google knows something about them (for example, their location, their age) but can only speculate how Google has ascertained this, and indeed how it uses this information. Like the epistemic uncertainties articulated by Ghostery participants in Chapter 4 regarding their algorithmic anticipation by data trackers, an epistemic insecurity is at work in these participants' speculations. However, instead of emerging as an *anxiety* in regard to how users are tracked, participants here display a sense of *trust* in Google to "know" and anticipate them. The "epistemic asymmetry" (Brunton and Nissenbaum, 2011) between data provider and data controller in this sense generates not concern that data tracking poses a threat to an inner and private self, but instead produces speculative faith that Google is capable of knowing users' identities and personalizing their experience to a complex and extensive degree.

Second, participants assume that the app is personalized to this high degree *even as the app gets these apparent predictions wrong*—for example,

Heena reported that she has no interest in stocks, but acknowledged that Google continues to display the card. Despite these failings, like Rachel, these users find personal relevance in Google's computational architecture. To return to the concept of technological self-validation, it appears that participants are reading *self-recognition* into the app—they find something "personal" in the app's inferences, despite the fact it is not anticipating them accurately. This kind of reading is not new and is not unique to algorithmic personalization systems: for example, as Adorno's ([1952] 1994) work on astrology illuminates, the reading of individual relevance into star signs involves a similar process of finding personal poignancy in texts that are in fact universally standardized and consumed.

In these algorithmically personal readings, however, there emerges an interrelation between text and reader that I argue is unique to the "predictive powers" of algorithmic personalization systems. Google's claims to "correct" inference are not *only* produced through a kind of epistemic faith in Google as some kind of "divine mind" (Hillis et al., 2013): after all, Google really does make use of some data in order to anticipate and infer the needs of web users (in a way that horoscope predictions do not). However, as established, this algorithmic anticipation is achieved largely though dividuating mechanisms: it is the individual user who must read relevance into such inferences as an individual. Google's predictive powers are therefore co-constituted by user and system: the algorithmic inferences deployed *inside* Google's algorithmic personalization technologies, combined with the user speculation of how such technologies work, create a kind of algorithmic faith that works to realize the "personalized experiences" promised by Google.

Finally, in regard to epistemic trust, the speculations of Lisa, Tariq, and Heena reflect what Bucher terms "the algorithmic imaginary": that is, "the way in which people imagine, perceive, and experience algorithms and what these imaginations make possible" (2016, 31). Their speculations of the power of Google reflect this algorithmic imaginary—an imaginary that emerges as epistemic trust in Google. More than this, though, participants' epistemic trust in Google can be considered again from an ontological perspective—participants' sense of self is not threatened by data tracking, as with the Ghostery participants interviewed in Chapter 4, but is instead *secured* in the speculations that Google "knows" these participants' preferences, identities, and even their parents. Algorithmic personalization's co-constitution of selfhood is not only a process that produces epistemic anxiety, or indeed a co-constitution that need revolve around protecting the inner self. In some instances, algorithmic co-constitution can be conversely considered as built on trust in the algorithm to secure and validate the self.

PERSONALIZATION VERSUS THE "IDEAL USER": GOOGLE'S NORMATIVE FRAMEWORK

In this section I explore the predictive power of the app, not through participant responses, but through an examination of the app's structural architecture. Such an examination is useful in providing an alternative explanation as to why Google might show a user "stocks" even if the user is not interested in them: namely, that the "stocks" card is delivered because it is one of a predefined selection of lifestyle choices upon which Google's predictive powers are predicated. As scholars such as Nakamura (2002), Oudoorshen et al. (2004), and Grosser (2014) have noted, the reliance on predefined, finite categories of lifestyle choice, interest preferences, and identity expression—such as the "stocks," "sports," "flights," "movies," "weather," and other cards on which the Google mobile app is built—have implications regarding "how data structures and computational power lead to certain kinds of interfaces or modes of presentation" (Grosser, 2014, n.p.).

These scholars propose that the predefined architectural and ideological frameworks that structure technologies can lead to standardizing, homogenizing, and reductive ideals of lifestyle choice and self-expression. Through such frameworks the "ideal user" of particular technologies, discussed in Chapter 3, is normatively assumed to represent the needs of "everybody," but in actuality is configured as white, male, heterosexual, and middle class (Oudshoorn et al., 2004). Take Google's "card" categories, which are used to frame and deliver the information that users "*need* throughout their day." Through an analysis of the cards themselves (rather than user engagements), it becomes apparent that Google is using a "limited choice interface model" (Grosser, 2014, n.p.) to construct and infer the lived trajectory of its assumed "ideal user" (Oudshoorn et al., 2004). For example, at the time of the study, the "sports" category was built on a set of structural protocols that *only* allowed users to receive updates about the pre-registered, exclusively male sports teams defined in Google's database, thus excluding the interests of users who might want updates regarding female sports teams. At the time of writing, however, the app has updated its software to include some women's teams—the limited choice of Google's interface has, it seems, been widened. Though undoubtedly a welcome update, it is interesting to note nonetheless that women's teams were added some time after the app launch, indicating that "male" sports teams were considered—and subsequently coded—as simply "sports teams," while women's teams were added as an afterthought. This, it could be argued, simply reflects the "reality" of sports culture in many countries worldwide, wherein male teams enjoy more interest and coverage than women's counterparts. However, even if such a normative assumption of cultural importance is used to rationalize Google's structural decision, the app reinforces dominant ideologies that perpetuate rather than challenge preexisting social hierarchies.

Even those categories that do not involve adherence to normative identity categories, such as "stocks," construct an "ideal user" in that they assume that "stocks" updates are of relevance to all potential Google users, despite the fact that stocks updates would most likely be of interest to only a small, affluent, and wealthy subset of the app's users. As such, Google's choice to show Heena stocks is far more likely to be predicated on Google's normative assumption—"a dominant narrative" reflective of the "hegemonic frameworks" (Noble, 2018, 24) that shape daily experience—that stocks information is "relevant" to the daily lived trajectories of its "ideal user." It is a very impersonalized framework of "ideal use" that here structures the "personalized suggestions" the app does and does not deliver, rather than the fact that Heena's dad is into investments.

The negotiations here between Google and these users reveal the app's operation via a homogenous, a-personal architecture, by offering cards that *appear* personalized yet adhere to a highly normative framework. However, Lisa's, Tariq's, and Heena's speculations do not entertain the idea that the app is simply *not personal enough*. In fact, they attribute Google's failures to provide "relevant" information to quite the opposite problem; it *does* know them personally (it knows Heena's parents are into stocks; it knows she is Malaysian and used to hot weather; it thinks Tariq is a "worker" because he lied about his age on his YouTube account), but has anticipated their preferences incorrectly. Once again, these individuals read something personal into the app, only this "algorithmic imaginary" (Bucher, 2016) actively reinforces Google's claim to enact personalization by explaining its failed predictions via imagined personalization technologies. The assumption here is that the app fails *because* it is personalized, not because it isn't.

All participants said that they had "no interest" in stocks, and though Tariq, Heena, and Lisa speculated that Google Now's choice to display "stocks" might be due to complex personalization processes, Laura and Giovanni (who were interviewed separately) answered differently. For example, when I asked Giovanni why the app shows him stocks even though (as reported) he has no interest in them, he answered:

> Because I think that . . . this is a product made to be suitable for everyone, so maybe I could be a business person, a businessman . . . or a normal person.

For Giovanni, then, the app is for "everyone," for a "normal person"—an identity category which is tellingly also conflated with being a "businessman" (a conflation also made by Google itself). I also asked Laura, "Why do you think the app . . . has options for things like sports and stocks?" to which she replied, "Probably just to support everyone's interests."

Giovanni and Laura, unlike Heena, Lisa, and Tariq, recognize that the display of categories such as "sports" and "stocks" can be explained by Google's imperative to provide information deemed "universal" or "normal" in everyday life. However, far from questioning this framework, Giovanni and Laura place *their* use of the app *outside* the boundaries of the "normal" usage expected of the app's ideal user: they believe that these categories are useful to "everybody," or "normal people," just not to them.

Rather than placing these students' lived experience outside the realms of "normal" use, I am inclined to approach their difficulties in finding relevance in the app through a focus on the disconnect between Google's ideal user and the everyday trajectories of these participants. I argue that the Google app consistently fails to live up to these individuals' expectations because the app's normative framework is too far removed from these students' lived experience: who do not "work," who do not "commute," who are not into "stocks," who have other ways to check the weather. Yet tellingly for these participants, Google's framework is not rejected as a normative idea of "what life should look like"—rather, it is reconciled through participants' acceptance that their lived experience must exist outside "the norm."

As Gillespie notes, computational categorizations—such as the sports, stocks, and commuter lifestyle categories upon which the Google app is structured—work to impose a powerful form of governance on the users exposed to and constructed by these categories. He writes:

> what the categories are, what belongs in a category, and who decides how to implement these categories in practice, are all powerful assertions about how things are and are supposed to be. (2014, 198)

Google's picture of "how things are and are supposed to be" thus constructs its users as stock market followers, jet-setters, workers, male sports team fans, and consumers, a picture of life that these participants did not recognize as relevant to their own experience. Yet, instead of questioning the a-personal nature of this apparently "personalized" system, participants (such as Tariq, Heena, and Lisa) put the app's flaws down to complex systems of algorithmic inference that lead to false prediction; others (such as Giovanni and Laura) positioned themselves as outside the "everyone" for which the app supposedly works.

The ideological implications of categorization are certainly not unique to Google: as the work of Noble (2018), Cohn (2019), Foucault ([1975] 1995), and Hacking (1986) stress, the kinds of identity categorizations privileged by socio-technical systems have their roots in long-standing hierarchies of difference that have for centuries been used to reinforce and legitimize existing structural inequalities. Noble (2018) argues that such hegemonic frameworks are increasingly reinforced and indeed compounded by commercial technological monopolies such as Google in their driving imperative to generate revenue. She argues that "search engine results reflect the values and norms of the company's commercial partners and advertisers and often reflect our lowest and most demeaning beliefs, because theses idea circulate so freely and so often that they are normalized and extremely profitable" (2018, 35–36). Furthermore, Finn (2017) proposes that Google's move toward lifestyle services means that, increasingly, Google's algorithms are not just deployed for informational search queries. Instead, as they find their place on users' mobile phones as intimate interlocutors in everyday engagement, "the algorithm is a medium for living, a pathway to experience" (2017, 76). The Google app therefore exemplifies the ways in which algorithmic personalization has the power to paint a normative picture of what life should look like: a life structured around an "ideal user" who is middle-class, an investor in stocks, a commuter, an affluent worker, and a jet-setter.

PARTICIPANTS AS MEDIA STUDIES SCHOLARS: LEGITIMIZING TRUST IN GOOGLE

In this section I want to devote some time to participants' status as developing media studies scholars. As mentioned earlier, all participants were enrolled in a digital media module, and as part of their academic program, participants studied the sociopolitical implications of data tracking and personalization in the week leading up to the last session of the study. This begged the question: How did their critical introduction to personalization intersect with their high expectations of the app?

Session Four of the study was largely designed to respond to any critical developments participants may have experienced. As part of their study that week, students had learned about data tracking as a potential invasion of privacy, as well as Pariser's (2011) argument that algorithmic personalization might create a restrictive, invisible "filter bubble" of consumption (as outlined in Chapter 2). I asked students how they felt about this critique. Tariq told me:

> Personalizing your internet experience is pretty bad because you're really just validing [*sic*] your own opinions, which is like, you don't really want to do that, you want to be exposed to a diversity of opinions.

Tariq here accepts Pariser's critique. Lisa, Rachel, and Heena similarly agreed that the effects of algorithmic personalization can be detrimental; however, they claimed that viewing personalized content did not negatively affect their own experience of the web. For example, Rachel explained:

> If I wasn't so interested in world events and stuff, and then I wouldn't get like maybe world events on the top of my thing [search results] um so some people aren't going to be as educated about that sort of stuff and I think that's quite important. But because I am, like it's not really a problem for me, I feel like I get all the stuff that I need.

Rachel seems confident that viewing personalized content is "not really a problem" for her because she feels she gets "all the stuff" she needs. If we take into account Pariser's assertions that we need to be aware of the "things we didn't know we didn't know" (2011a, n.p.), however, Rachel's point becomes slightly problematic; in his theoretical critique, personalization algorithms prevent us from exposure to information that we cannot know exists (Pariser, 2011). Similarly, as mentioned as part of the data-for-services exchange discussed earlier, Laura displayed a critical awareness that Google might "manipulate" her, but she legitimized her acceptance of this through an individualistic rationalization that she is aware of such manipulations. In these responses, students' critical engagements with personalization are weighed as applicable to other people, *but not to them*. Thus, their negotiations with Google as a trustworthy sociotechnical force are legitimized through a kind of critical distance between "other people" and the participants themselves.

Conversely, however, in Session Four, Giovanni told me that the module had made him "more aware" of privacy issues, and suggested that the only way in which he was going to use the app was as a form of resistance *against* Google's data-tracking strategies:

> *Interviewer: Do you think you're going to continue using Google Now?*
> GIOVANNI: Er, I don't know, I mean I think I'm going to use it randomly, like when I have to search for something on Google, just to for piece of mind to check what Google Now is saying.
> *Interviewer: When you say for piece of mind, what do you mean?*
> GIOVANNI: To check that Google Now hasn't taken too much personal stuff.

This resistant use exemplifies Gillespie's statement that "[w]hile it is crucial to consider the ways algorithmic tools shape our encounters with information, we should not imply that users are under the sway of these tools. The reality is more complicated and more intimate" (2014, 186). However, it seems important to acknowledge that throughout the study, Giovanni's trust in Google was muted; therefore his critical development largely remained consistent with his engagement with the app. For the other participants, however, though they were aware of the potential detrimental effects of algorithmic personalization, algorithmically personalized information was to be embraced as convenient and largely beneficial to these individuals, even if there were wider concerns for more universal use. In some ways, such responses reflect those of the Ghostery users who used tracker blockers to protect themselves at the expense of other individuals. Once again, I am reminded of Brunton and Nissenbaum's statement that in the face of the "unknown unknowns" (2015) that datafication creates, web users must make do with whatever logics they can work with, and through, in order to make sense of epistemologically unknowable algorithmic infrastructures in which they find themselves entangled.

CONCLUSION: "I'LL USE GOOGLE, JUST BECAUSE IT'S THERE NOW"

In the final session, I asked participants to sum up their experiences of the Google app. Overall the response was muted: Lisa stated, "it promises a lot, and it sort of didn't really help me in any way . . . its not as impressive as I thought it would have been," though later in the session she invoked an element of self-blame, stating, "I mean I'm not sure if that's my fault for not giving it enough information." Similarly, Rachel said:

> I'd just say it was disappointing . . . I just, I think I expected a lot from it, like from, what's its slogan? Like giving you the information you need before you even ask for it, well that's what I expected . . . but I barely got any information on it when I asked, so [shrugs].

For almost all the participants, it transpired that the app had in general failed to live up to their expectations. To conclude, I reconsider the questions asked at the beginning of this chapter: In the face of this failure to live up to expectations, are these users' lived experiences informed, altered, or constructed by Google's "personalized" suggestions? Were their lived trajectories redirected or reconstituted in any way by the app's inferences?

In one sense, the answer is a fairly unequivocal "no": given that the study's participants repeatedly failed to find a use for the app, their lived experiences were almost inevitably left uninformed and untouched by Google's predictive powers. It seems that the app's very disfunctionality has rendered its autonomous interventions somewhat redundant: Google is not acting on, constructing, or altering the lived experiences of these participants in ways that Facebook's apps intervened in users' identity performances, as detailed in Chapter 5. In some ways, then, failure to find a use for the app, coupled with the disconnect between Google's "ideal user" and their own lived experiences, seems to have formed a kind of unintentional resistance to any tangible forms of governance Google might have imposed on their lived trajectories. To put it more crudely, these participants were not put "under the sway" (Gillespie, 2014, 184) of the app's predictive powers because the gap between the "anticipated user" and the "user herself" (Gillespie, 2014) is simply too great.

I do not mean to suggest that is this gap were closed—if Google *did* manage to align its predictions with the lifestyle trajectories of these users—then these students' lived trajectories would be necessarily constituted or conditioned by Google. There are many ways that users resist, appropriate, or turn to face the technologies entangled in everyday experience, as the work of Kennedy et al. (2015), Cohn (2019), and Bucher (2016) highlights. However, given that the students could not find a use for the app, I do not feel justified in critiquing how these participants' usage could be read as regulatory or resistant. Nor do I mean that if Google's inferences did "work properly" then they would be discounted from critical interrogation—as Hall (1989), Jarrett (2014a), Deleuze (1992), and other media and cultural theorists stress, the structural mechanisms of neoliberal capitalism mean that even as these structures may afford benefits, they can still work to regulate and discipline the subject. What I do propose, though, is that even as this disconnect between "ideal user" and these participants' lived trajectories arose time and time again, the participants were happy to place their trust in the app's predictive and personalizing *promises*. This trust was not based on the app's functionality. Instead, it emerges from participants' adherence to discursive normative frameworks that are implicit in Google's contemporary drives to personalize users' experience. Participants assumed that the app *must* be convenient, that is, it must know them and it must be able to personalize for them, even though it failed to function in this manner.

The epistemic trust placed in Google by these users emerges from Google's power to validate the self through both algorithmic inferences deployed "inside" the algorithm and the "algorithmic imaginary" (Bucher,

2016) brought to the engagement by the users themselves. It is therefore the relation between user and system which *co-constitutes* Google's uneasy entanglement in these users' lived experiences. By this I mean that participants *worked hard* to find something personal in Google Now's normative framework—from Rachel's pleasure at recognition, to Heena, Lisa, and Tariq's assumption that the app is capable of knowing their age, their nationality, and who their parents are. Though participants' lives in some ways remain unconditioned by the limited functionality of the app, the predictive power of Google remains epistemologically unknowable but also unquestioned. It is therefore the *possibilities of personalization*, and not its tangible functional outputs or uses, upon which participants' trust and pleasure in Google is predicated.

This embrace was exemplified in Rachel's concluding remarks of her engagement with the app. Though Rachel described her experiences during the study as disappointing, she later implied that the app's very existence was enough to reorient her practices to incorporate its functions, as evidenced in the following exchange:

> RACHEL: I'll probably still use [the app], if I like need to find somewhere in [the local town], I can always like use it to find stuff, so I probably still will.
>
> LISA: But you can just use normal Safari for that can't you?
>
> RACHEL: Yes, you can [laughs]. But I'll use Google, just because it's there now.

Remember, Rachel was the participant who was most consistently disappointed in the app, but who enjoyed the moment of self-recognition that Google Ad Settings awarded her. In her attempt to find a place in her lived trajectory for the app, Rachel's statement highlights Google's power as a monolithic force. Its ubiquitous and naturalized status in everyday life, combined with the fact that Google can add its personalization services to users' devices automatically and free of charge, means that Google no longer needs to "work" in order to create the opportunity to attract and keep users. As discussed, participants found a way to "make personal" Google's distinctively a-personal operations, even as they failed. Google's predictive promises, combined with its naturalized presence, become the reason in and of itself to trust in Google as a socio-technical force, a predictor of the "best" everyday trajectories, and even as a stabilizer of selfhood. As Rachel says, she'll use the app, "just because it's there now."

Conclusion

Removing the "Personal" from Personalization

In their book *Reinventing Capitalism in the Age of Big Data*, Mayer-Schönberger and Ramge (2018) paint an optimistic vision of a datafied future, one where data-driven exchanges have superseded conventional money-based transactions as society's core system of value measurement. They note that Big Data has all of the characteristics of conventional money in terms of its transactional and standardizing qualities, but that data are also able to measure and make exchangeable *individual* registers of value: for instance, an individual's particular want, preference, or desire for a specific product or service. For these scholars, Big Data will facilitate optimal socioeconomic market transactions, noting that in the near future "we'll combine huge volumes of such data with machine learning and cutting-edge matching algorithms to create an adaptive system that can identify the best possible transaction partner on the market" (2018, 4–5).

Mayer-Schönberger and Ramge (2018) are not the first to compare Big Data to resources and value systems currently fundamental to society: data have been hailed as the "new gold" (Peters, 2012), or critiqued in their discursive construction as a "raw material" (König, 2017) despite

Making It Personal. Tanya Kant, Oxford University Press (2020). © Oxford University Press.
DOI: 10.1093/oso/9780190905088.001.0001

their very human-engineered constitution. However, Mayer-Schönberger and Ramge's (2018) vision of datafied markets is perhaps one of the most optimistic celebrations: they are hopeful that data will offer more consumer choice, more "satisfying work," and through "data-richness" create a future that "is going to be profoundly social and thus deeply human" (2018, 223). Such an optimistic vision can of course only ever be speculative; however, I propose that in some ways a problematic version of their datafied future can be considered already here. As this book has sought to establish, on the contemporary web, users are increasingly compelled to relinquish their data in exchange for access to cost-free (in monetary terms at least) content and services: data already function as a key form of transactional exchange. More than this, though, this data-for-services exchange relies on algorithmic personalization technologies to generate value: as Mayer-Schönberger and Ramge's speculations suggest, Big Data is *only* valuable because of the adaptive, decision-making systems that anticipate the identities, desires, and needs of social subjects in order to individually target them.

This current data-for-services exchange so common on the contemporary web may well lay some of the foundations for a datafied future. However, though I share Mayer-Schönberger and Ramge's (2018) emphasis on personalization systems as holding a key function in the data economy, my focus is less on the speculative and macrocosmic opportunities that algorithmic anticipation might one day provide. Though such opportunities might generate new forms of free-market capitalism, they do less to generate social value for the data subjects implicated "in the maximization of profit" that Mayer-Schönberger and Ramge (2018, 220) are eager to retain. I am interested more in the situated particularities of the datafied present—the engagements and entanglements that algorithmic anticipation creates for users on the contemporary web. As the accounts of web users featured in this book highlight, there are some key problematics that emerge in users' encounters with algorithmic personalization. These accounts come together to suggest a number of critical issues with market-driven algorithmic personalization as experienced in everyday life.

The first is that of knowledge production. The tracking of users certainly does generate more knowledge of users' identity categorizations, likes, and preferences—albeit in epistemically asymmetrical ways (Nissenbaum and Brunton, 2015) that afford far more insight to platforms than data subjects. Powerful platform providers have little interest in "who the user is" as a social subject but rather look to identify the subject as a consumer and dividuate them into commodifiable categories. Though data can be used to process and generate knowledge about web users *for* web platforms and data brokers (Mai, 2016; Brunton and Nissenbaum, 2015), my research indicates that the epistemic opportunities afforded by data *to web users themselves* are tense, nuanced, and complex.

The epistemic asymmetries between users and platforms have drawn calls for more user "awareness" (Willson and Leaver, 2015; McStay, 2017) over how data are managed and processed. These calls come in response to the imbalance between platforms and data subjects that no doubt contributes to the epistemic uncertainty felt by web users—yet I am reluctant to join them. Increasingly, as is illuminated by the participant testimonies contained in this book, web users *are aware* that they give up their data in order to access and use cost-free web services. The Ghostery, Facebook, and Google app users who gave their time and testimonies to this research indicate that the provision of data in exchange for free services is rarely undertaken in complete ignorance from the user's perspective. This provision of data is articulated through different discursive legitimizations: for the Ghostery users interviewed, being subject to dataveillance was strongly resisted, or at most reluctantly accepted as long as the right exchange in data-for-services was negotiated with specific platforms. Such reluctant resignations echo findings by Turow et al. (2015) and Ofcom (2019) that users feel resigned to giving up their data under conditions they feel they have little control over. As such, consent to data tracking might be reluctant, but it would seem that increasingly web users are not "ignorant" of their positions as data providers. In participant accounts of reluctance and rejection, I argue that the uncertainties inherent in data tracking emerge as *epistemic anxiety* over the ways in which we are known, anticipated, and possibly exploited by data trackers. However, my findings suggest that anxiety over data tracking is only one avenue of negotiation that web users take in their engagements with data. For other participants, such as the Google app users interviewed, data provision was not rejected but rather embraced as the right price to pay for personalized services. In such engagements, epistemic uncertainty emerges as *trust* in algorithmic personalization systems—a faith that platform providers will uphold their side of the bargain in the data-for-services exchange. The testimonies of the users in this book highlight that the boundary between ignorance and awareness should not be considered in binaric terms: though users are aware that they *are* being tracked, the specifics of how, when, and why remain an epistemic uncertainty that is navigated through a plethora of situated responses.

The findings presented in this book suggest that the issue of epistemic uncertainty is not one to be "solved" through an increase of access to knowledge about third-party tracking. As I explored in Chapter 4, the epistemic uncertainties of being made in data (Cheney-Lippold, 2017) actually *increased* for those web users interested in protecting their privacy: the more knowledgeable the web user, the less able they felt to stay "on top of" their data, as interviewee Robikifi put it. As such, I would argue that to call partial awareness of data tracking "ignorance" is to wholly misrepresent the epistemic impossibility of coming to grips with the entirety, specificity, consequences, and effects of your own data trail. There is no "getting on top" of your data because essentially, there is no "top": as Mai (2016) argues, data mining does not just unlock, manage, and concretize existing forms of knowledge, it creates new knowledge about the future and, as Cheney-Lippold (2017) stresses, about the present. The reception of the EU GDPR law is a case in point: numerous news and popular reports have highlighted that (1) the public feel entirely overwhelmed (Kelion, 2018) by the endless requests for consent that the GDPR creates, and (2) that the implementation of the law by data trackers means that new forms of knowledge are still perfectly capable of being produced about data providers in ways that are incomprehensible to data subjects themselves.

I am similarly reluctant to join calls for more "control" over data tracking. In the European Union the aforementioned GDPR laws seem to have taken a step forward toward giving users more control, but research is already emerging which suggests that data brokers are finding ways to circumvent the rejection of tracking by users (Kiskis, 2018). The rollout of the Open Banking Law—which allows platforms such as Facebook access to consenting EU consumers' financial and bank account data (Hernæs, 2017)—similarly suggests that even as cookie notices appear to be giving users "more control" on an individualized level, from a macrocosmic market perspective legislative changes are actually opening new ways to track, manage, and anticipate users through not just their social interactions, but their financial data. To call for more user awareness and control therefore does not seem to lead to effective ways to assert user and public autonomy; if anything, it simply prompts a neoliberal solution to a condition legitimized by neoliberal rhetoric. As I explored in Chapter 3, algorithmic anticipation is largely enacted under discourses that champion "you" the user as a sovereign subject and agent of your own success. However, in being positioned as individually responsible for their own dividualized data trail, users are asked to take on the impossible burden of knowing and consenting to epistemologically unknowable data management.

FROM "PERSONALIZED SEARCH" TO "SEARCH": DISCURSIVE ERASURE

Algorithmic personalization practices, like many web-based technologies and operations, are likely to become more developed as web users adapt their interactions to embrace or reject the data-for-services exchange so common on the web, and as data tracking becomes increasingly ubiquitous and advanced. It is telling, for instance, that developments in personalization also include discursive erasure of the term itself. For example, Google no longer employs the term "Personalized Search" (Google Blog, 2009) to describe the tailoring of search results based on individual "relevance." This is certainly not because Search is becoming once again more "universal" and "objective," as it was when it was first launched (Van Couvering, 2007)—as Google's information videos continue to make explicit, Google's "goal" is still to "create a seamless connection between [users'] thoughts and their information needs and the results they find" (Google, 2016). The desire to anticipate the individual in the name of convenience (and of course commerce) still persists, but the term "personalized" does not feature in any of Google's current information materials. The paradox of this, however, is that this apparent "personalization" of search actually works to reinforce sociocultural hierarchies of normative identity and lifestyle. As Noble (2018) asserts, Google search is built on dominant ideologies of race, gender, and class that mean Google's search results continue to marginalize social subjectivities while they uphold the "ideal user" as male, white, and US-centric. My analysis of the Google mobile app found a similar homogenizing framework in Google's so-called personalized inferences; as I established in Chapter 6, despite Google's predictive promises, their mobile app paints a normative picture of what life should look like. And yet in these impersonal mechanics, the Google users I interviewed sought to read individual relevance into the Google mobile app's interface. These users found something personal in Google's impersonal predictions, even as they repeatedly failed to find a use for the app.

The a-personal mechanics of platforms like Google might beg the question: How can they still be considered personalization technologies? As I have established throughout this book, despite the impersonal nature of personalization, platforms continue to data mine and anticipate users in ways that *do* intervene and reconstitute their experiences. These interventions are based little on Google's and Facebook's interest in users as holistic social subjects: they are more interested in disassembling and categorizing users into commodifiable, dividualized data sets (Bolin and Andersson Schwarz, 2015; Cheney-Lippold, 2017). The term "personalization" continues to be discursively deployed by platforms to legitimize the dividualizing process on which data trackers rely to generate revenue, albeit only in the cookie notices and privacy policies users now so commonly must consent to in order to use platforms' free services.

Such omissions suggest not that personalization is disappearing, but that the notion that platforms should track us, anticipate us, and *act* on us is becoming ever more ubiquitous and naturalized: what was once "Personalized Search" is now simply "Search." In fact, if anything, the constant offers of increased "knowledge" experienced by web users in the form of consent notices is helping to normalize the practice of personalization. The most paradoxical example of this discursive normalization can, I argue, be found in the rollout of the aforementioned GDPR privacy legislations in the European Union. Such legislative measures, as I have established, seem to do little to enlighten users in ways that afford users more autonomy as data subjects. In establishing "consent" to data tracking and algorithmic anticipation as an inescapable part of web experience, the GDPR seems to be actually working to concretize the very practices it is supposed to be demystifying. Data tracking, conducted in the name of personalization, is becoming something of an "open secret" on the contemporary web.

Westling (2020) highlights that despite its apparent naturalization and ubiquity, anticipating audiences via their identity markers is not and should not be the only way that platforms can understand their audiences. She makes a compelling case that platforms should epistemologically understand web users not through their status as identity-marked, singular, individual *agents*, but instead through their (collective) dynamic *agencies*. Noble (2018) and Cohn (2019) remind us, however, that the apparent disembodiment of agency should be carefully acknowledged as still something entangled with and within structures that adhere to dominant hierarchies of gender, race, and class, even when the agent is taken to be "universal." Furthermore, such an approach involves platforms understanding their audiences as dynamic flows, rather than commodifiable objects—an approach that seems unlikely to take hold any time soon in a web economy that relies so heavily on the idea that identity markers such as age and gender should matter so much to the personalization process.

As Chapter 2 highlighted, it is within the context of this commercial drive to personalize that mapping the negotiations of web users becomes pivotal. There is a pressing need to continue to critically consider the *ways* in which users negotiate, understand, and are entangled with(in) personalization practices if we are to understand what discursively naturalized anticipation systems do to the users they anticipate. For those web users who currently encounter algorithmic personalization as part of their current lived experiences and trajectories, the drive to personalize demands negotiations for autonomy, identity, and epistemic knowledge production.

STRUGGLES FOR AUTONOMY

It is not only the epistemic uncertainties inherent in data tracking that have provided a focus for this book. Though utterly invaluable for understanding how data have material and performative implications for identity and beyond, I would argue that to focus only on *back-end* data-tracking practices obscures the front-end interventions that algorithmic personalization makes in users' digital engagements. Exploring identity and everyday life in the context of the contemporary data economy must therefore also take seriously the role that algorithmic personalization technologies play in reconstituting the experiences of users as they scroll, click, and inhabit their daily networked trajectories at the level of the interface.

Once again, Mayer-Schönberger and Ramge (2018) work highlights the pivotal place that "adaptive systems" play in the data economy. As these authors note, adaptive systems will benefit individuals for their powers in algorithmically personalizing web user engagements with what are at present unmanageable amounts of data, services, products, and content. Adaptive systems, they predict, will be able to preempt user preference in order to manage information streams, to filter out content and services irrelevant to individual users, to suggest the best products—not based on price, but based on their specific desires and needs—and indeed to make everyday decisions in your stead. I would argue that such adaptive systems already exist in numerable but crude forms all over the web: the examples of autoposting apps and Google's mobile app that underpin the empirical research in this work are both forms of personalization that seek to "streamline" user experience and save them time by algorithmically acting on and for them. Such algorithmic decision-making systems are still very much in their infancy, but Mayer-Schönberger and Ramge predict that one day they "could lead to systems that come already preloaded with a robust, comprehensible set of preferences—a smart, even-keeled decision agent that can step in whenever you do not trust your own judgement" (2018, 80).

Mayer-Schönberger and Ramge (2018) seem to stipulate that it will be an easy task to determine which decisions we want made by personalization algorithms—those that save time, or those which are boring, or even choices we find difficult because of our "biases." According to these authors, our entanglement with the algorithms that act in our stead will be an uncomplicated dynamic in which we will act as commander and our personalization algorithms simply as assistant. However, as the testimonies analyzed in this book have illuminated, even with the present and somewhat crude decision-making algorithms users currently encounter, the line between user and system as distinctive autonomous agent is a hard one to draw, and an even harder one to navigate, at least from the user's perspective.

For the web users interviewed for this book, the realities of negotiating with algorithms that act with, for, and on us are never simple. Web users mobilize a number of tactical maneuvers in their encounters with these systems: for the Facebook users whose identities were (re)written through autoposting apps, personalization algorithms did not just "speak" for them but intervened and disrupted their identity performances in frustrating, "embarrassing," and difficult ways. For others, algorithmic tactics were employed, and not always enacted to take back control of self-performance but to actively turn to face the algorithm: to reorient their activities to suit the protocols of algorithmic logic (Gillespie, 201, 184). In the context of these negotiations, the decision-making capacities that Mayer-Schönberger and Ramge (2018) celebrate as the future emerge as present struggles for autonomy between user and system in everyday contexts. These struggles are sometimes mundane and are not always enacted as resistance, but they nonetheless indicate that the increasing outsourcing of decision-making to personalization systems has complex implications for identity, autonomy, and knowledge production that extend far beyond deliberations of time-saving and convenience.

I want to stress here, as I did when introducing the empirical data that constitute much of this book, that participant accounts analyzed in this book are context, time, and subject specific: they illuminate a historical moment in which the commercial drive to personalize is in its technological infancy and is yet an overarching discursive framework that continues to be used to legitimize the ubiquitous tracking and anticipation of web users. In historically contextualizing the drive to personalize as both a techno-economic development and as a neoliberal discourse, it becomes possible to consider the "next steps" in the trajectory of the commercial web. For instance, the struggle for autonomy described throughout this book is still being played out in the functionalities of digital "smart speakers" such as Google Home and Amazon Echo. These developing algorithmic assistants actually seem to have taken a step *away* from their status as decision-making agents. Instead of "giving you the information that you need before you even ask" (Google, 2014), as the Google app has emphasized in the past, these assistants are more marketed on their voice-recognition abilities to literally take commands from the user in the form of "Alexa, open Spotify for me." In such user- rather than system-initiated engagements, the lure of algorithmically inferring a user's needs seems to be giving way to personalization systems that are a little less sure of "what you want." Perhaps the public is ever more reluctant to relinquish too much autonomy to personalization systems in the quest to make life easier. Of course, to use Alexa, users have to reorient themselves toward its technological capabilities—to say the right things and feed it the right information. Users must literally turn to face these algorithmic assistants, even if the decision-making capacities are not as pronounced. Ironically, in turning to face these algorithms, web users are further reinforcing their positions as "data providers" for commercial systems that look to marketize and monetize social interactions, even inside the home.

The rise of voice-command personal technologies might be explained through users' reluctance to be spoken *for* by algorithms and instead to speak *to* them. However, this development might be platform- rather than user- driven: voice-recognition dominates because at present it is technologically more developed than other forms of technology that lend themselves more to algorithmic control. Like all entanglements, the "root" of the rise of voice-command is messy and involves multiple socio-technical and political economic actors. This brings me to an important point: the struggle for autonomy between user and system, when explored through lived experience, is revealed to be a struggle not between two easily distinguishable actors, but a complex negotiation between different platform providers, a multitude of technologies, other social subjects within a network, neoliberal ideologies, the "algorithmic imaginary" of users (Bucher, 2016) and the "algorithmic imagination" of the system (Finn, 2017). I am reminded here of participant Beth's negotiation with her Spotify-Facebook profile connection, which entangled a multitude of actors and actants: her Facebook friends' network, her public but in fact very much private performance of listening to music while walking down the street, the definitions of "public" and "private" imposed on Beth by Spotify, and the algorithmic agents that position songs as markers of identity, but also as advertising opportunities. The algorithmic entanglement of user and system created by personalization is an entanglement not between two agents, but among a multitude of actants that constitute lived experience.

MAKING THE SELF: REGIMES OF ANTICIPATION?

The agential capacities of personalization algorithms means that algorithmic personalization emerges as one of many "force relations" (Bucher, 2016) that intervene in and co-constitute the daily experiences of web users. These force relations are far from unique to the computational in establishing the horizons of possibility in everyday life: self and social world are always subject to a process of constitution in, through, and with the political, cultural, and economic frameworks that present historically specific possibilities for living. However, the disciplinary power of algorithms has come to play an increasingly significant role in digital engagements that now extend far beyond the computer interface: combined with the global flow of capitalism, algorithms have the disciplinary potential to prompt a reconsideration of the ways in which the self is structured and maintained in twenty-first-century cultures. The detriments of being constituted in and through performative algorithms are material, profound, and unequally distributed: Hearn (2016), Skeggs (2017), Chun (2015), Cheney-Lippold (2017), and Noble (2018) especially stress the alarming consequences of being categorized in and through data. Those deemed valuable by algorithms are given the chance to reinforce their social status outside of the algorithm, while those deemed computationally less valuable can be targeted financially, marginalized, or indeed rendered invisible on social networks.

This book has made a strong case for the crucial role of the everyday in understanding how largely theoretical models of algorithmic governance play out in the users' lived experiences. The accounts of the users interviewed for this research suggest to me that to be made and valued in data is *not* simply a process enacted from inside the database, or from within the process of algorithmic constitution. As Bucher asserts, the ways in which we are performatively constituted also involve the ways in which users see themselves "through the 'eyes' of the algorithm" (2016, 35). The constitution of algorithmic activity emerged again and again in participants' accounts, but perhaps most especially in those featured in Chapter 6, wherein interviewees "redrew the boundaries" (Burns, 2015) of the Google app as socio-technical object by *reading into* the app a predictive power that the app itself failed to display. I argue that users find themselves embedded within regimes of algorithmic anticipation (Hearn, 2017) that structure the everyday, but that users themselves also help to *co*-constitute.

In approaching algorithmic anticipation from the perspective of the users themselves, algorithmic personalization emerges as not only an epistemic but also an *ontological* concern: the ways in which individuals are "known" by algorithms have performative implications for users' sense of self. For Ghostery users, algorithmic anticipation posed a threat to an inner self, one that preexists the network and that must be protected from the dehumanizing gaze of data trackers. The self here corresponds to an idea of identity ironically echoed in Zuckerberg's well-quoted claim that users "have one identity" (cited in Van Dijck, 2013). Unlike Zuckerberg's "one identity," that can and should be expressed in all its glory through social media, Ghostery users conceptualize a self that should instead *be protected from* the network. Current rhetorics that celebrate algorithmic personalization reinforce this idea of the self as unitary and authentic: the self must also correspond to an "authentic self." This is a market imperative: to commodify datafied social interactions, advertisers, marketers, and other data brokers must know what a user is "really" interested in.

However, in the quest for "truly" knowing users in "all" of their marketable subjectivity, a strange contradiction emerges: users are asked to authenticate a coherent, fixable self, yet also iteratively and abundantly express themselves in all ways across all platforms (Szulc, 2018). Data brokers' encouragement of endless self-expression by users—to register and publicize their "likes," their locations, "their current thoughts," and "feelings"—is driven by the need to grammatize such expressions for the algorithms tasked with making sense of "all the things you do." In doing so, another formation of self emerges: as I argued in Chapter 5, this selfhood is one that can be expected to perform identity across all networks and further permit algorithms to promote and manage not just data, but identity expression itself. The result for the user is a self that, via context collapse (Marwick and boyd, 2011), is foreclosed of the possibilities to enact different identity performances for particular social groups in specific settings. More pressingly, it is a self that has the paradoxical quality of being brought into existence via algorithmic actors that removed the subject as the key author of identity expression. In providing algorithmically "personalized" utterances of identity for users, the users are performatively brought into existence at the very moment they are written out of their self-performance.

The formations of selfhood described so far perhaps paint a rather bleak picture of only *unwanted* entanglements of users with algorithmic personalization. The Facebook and Ghostery users who gave their accounts to this research expressed a resistance, or at least reluctance, in being acted on and for by anticipatory data-tracking and algorithmic mechanisms. However, the engagements of Google app users explored in Chapter 6 offer an alternative perspective to research that frames technology use through resistance or rejection. In this case, participants looked to work *with* Google's personalization systems in an attempt to get more out of their use, and indeed to find something personal in Google's overwhelmingly impersonal idea of what life should look like. These participants looked not to turn away from algorithmic grammatization, but instead to adhere to the structural necessities demanded by Google. In the face of epistemic uncertainty, it is not anxiety that drives web user response, but faith: an epistemic trust in Google to personalize their experiences to an extraordinary degree. In this tactical maneuver toward algorithmic power comes a formation of selfhood that is validated in algorithmic anticipation: the self as social subject is legitimized as existent not in the eyes of other social subjects, but instead in and through data.

FROM UNDERSTANDING TO COPING: DATA PROVIDERS AS ALGORITHMIC TACTICIANS

The accounts and testimonies given by web users to this research do not cover but certainly illuminate the myriad of ways in which web users position themselves in relation to algorithmic personalization. Personalization algorithms in the everyday both "take something" from social subjects in the form of data, but also "give something" back: a personalized ad, a relevant book recommendation, an autofill search query, a suggested commute to work, or even an offer for a credit card that they can't afford. It is the reported front-end experiences articulated by research participants that have provoked a re-evaluation of my questioning of how users "understand" algorithmic personalization. I stated in Chapter 1 that algorithmic personalization is "slippery"—it is both there and not there, "felt" by users and legitimized by platforms but not pinned down. This book has explored not only how users *understand* algorithmic personalization, but how users "cope" with the fact that as data providers, we do not and cannot fully understand all the ways and means in which platforms seek to anticipate and act on our movements and interactions.

I do not mean to suggest that in "coping" with algorithmic personalization that its effects are somehow only "felt" or "imagined" by those that encounter it. Quite the contrary: the operations of personalization are tangible and material—it is just that for those web users who encounter such practices, these practices remain elusive but effective. I mean "effective" not in the sense that they "work" (for some participants in this book, algorithmic personalization did not "work" for them), but in the sense that they have tangible effects on everyday experiences with online and web technologies. The computational operations deployed to make personalization possible may be slippery to those subjected to them—but they continue to tangibly intervene in user experiences and everyday trajectories nonetheless.

At first there might seem to be connotations of powerlessness in the idea of "coping" with algorithmic personalization: the term is suggestive of a state that individuals inhabit *in reaction* to something, a position of having to "deal" with a situation rather than actively lead it. However, I would argue that coping best encompasses the ways in which web users engage with algorithmic personalization systems. This is because, though web users can and do negotiate, turn toward or turn against, put their trust in or remain suspicious of the systems that seek to know them, in all of these engagements web users are first and foremost positioned as data providers: a position they must, in some way, respond to. Even if individuals respond by not "accepting" the terms of service that promise them a personalized experience in exchange for their data, the epistemic uncertainty inherent in algorithmic anticipation renders trust in such terms nearly impossible. How, why, and when we engage with algorithmic anticipation is debatable and does open up opportunities for users to deploy a range of tactical maneuvers, but our entanglement with algorithmic personalization is non-negotiable: it is a market driven pre-condition of the digital everyday.

It is in this idea of coping that I want to conclude this book—not coping as a passive response, but as what I call an algorithmic tactic. Algorithmic tactics do not refer to the functions and possibilities deployed by personalization technologies: to draw once again on De Certeau's (1984) framework, algorithmic anticipation should be considered a strategic maneuver implemented by epistemically powerful platform providers and data brokers. Instead, algorithmic tactics refer to the ways in which users themselves maneuver within, against, and through algorithmic anticipation. It is in user negotiations with algorithms that questions regarding epistemic engagement begin to emerge as productive. As aforementioned, calls for increased user "knowledge" or "control" are unhelpful; it is not an evaluation of knowledge we need—it is something more akin to a evaluation of tactical engagement. By this I mean that in our individualized orientations toward algorithmic personalization, being able to deploy certain types of tactical play to resist, turn toward and outmaneouvre anticipation algorithms is what matters, not "further insight" into data tracking.

To use a term proposed first in Chapter 5, I argue that the deployment of *algorithmic capital* might function as a key framework in understanding the epistemic and autonomous struggles of being algorithmically anticipated by platform providers and data brokers. In Chapter 5 I argued that algorithmic capital could be deployed or accrued in ways that legitimize one's own orientation toward the algorithm as "proper" while delegitimizing others' engagement with algorithms as illegitimate. As autonomous algorithms solidify their place in everyday life, there will increasingly be "right" and "wrong" ways to cope with algorithmic intervention. Of course, the mobilization of algorithmic capital to judge, value, or devalue *other* web users cannot and should not be considered some kind of social good: like all forms of cultural capital, the display and accruement of algorithmic capital functions to indicate intrinsic worth in ways that mask the culturally constructed nature of social distinction. More simply put, there is no "right way" of engaging with the algorithm, at least in regard to the valuing of user-to-user interaction.

Though algorithmic capital is an inherently problematic mechanism for the valuing of social subjects, it *might* be a useful alternative to understanding algorithms without resorting to calls for "more knowledge," "more control." We need to deploy algorithmic tactics not against other users but toward platform providers and data brokers, in an effort to understand how we orient ourselves toward or against algorithmic entanglement. As De Certeau argues, everyday tactics are situated responses that cannot give social subjects "a view to the whole" (1984, 39). But they can offer some partial re-affordance of autonomy within structural frameworks that place platform providers and data harvesters in strategically powerful positions. In recognizing these maneuvers as less powerful, data providers can and should be acknowledged not as agents of their own (datafied) success, but as algorithmic tacticians who "make do" in their everyday digital trajectories in meaningful, creative, and productive ways. The affordances of web users as algorithmic tacticians cannot be taken as standardizable or universal: the burdens of being algorithmically constituted are unequally distributed and felt. Such inequalities preexist algorithmic constitution, but are perpetuated and redrawn by algorithmic anticipation in ways that mean we must pay critical attention to the difficulties that such anticipation imposes on particular groups of social subjects. As in Chapter 1, I once again call for further qualitative work that attends to the algorithmic anticipation of specifically situated and marginalized social subjects.

The investigations that constituted this research have in many ways highlighted that identity is (always already) collectively informed and socially legitimized. However, they have also uncovered new forms of algorithmic entanglement between users' sense of self not only as unitary, private, and interior, but also as multiple, context specific, and recursively reworkable in ways that set up algorithmic personalization as a legitimizing performative power. My research has found that it is users, as well as platform providers, that constitute the self in relation to everyday algorithmic anticipation. Under the complex yet "black boxed" ways that algorithms seek to know us, "understanding" the algorithm becomes less important than negotiating the interventions of algorithmic personalization into everyday life. As user anticipation becomes more ubiquitous, the ways in which the self is constructed is increasingly entangled with the algorithm. It is through lived experience that the complexities of being made in, and making do with, algorithmic personalization practices, that such practices are revealed as performatively (re)constituting users' knowledge, autonomy and the self.

APPENDIX

1. INTERVIEWS FOR GHOSTERY STUDY

TABLE A.1

Participant Information (Agreed Pseudonym, Occupation, Country of Residence)	**Interview Date**	**Format**
Katherine, managing director, Netherlands	9/25/2013	Face-to-face (audio recorded)
Edward, occupation undisclosed, France	9/26/2013	Email
Christopher, senior systems engineer, US	10/5/2013	Email
Gyrogearsloose, unemployed, Canada	10/30/2013	Skype (audio recorded)
Participant, undisclosed, undisclosed	11/10/2013	Email
Mary, web developer, US	11/9/2013	Skype (audio recorded)
HelloKitty, unemployed, UK	11/12/2013	Face-to-face (audio recorded) *Note: interviewed with Yoda, HelloKitty's partner*
Yoda, IT user support officer, UK	11/12/2013	Face-to-face (audio recorded) *Note: interviewed with HelloKitty, Yoda's partner*
Robkifi, researcher, UK	11/14/2013	Face-to-face (audio recorded)

Participant Information (Agreed Pseudonym, Occupation, Country of Residence)	Interview Date	Format
Claire, postgraduate student, UK	11/15/2013	Face-to-face (audio recorded)
Lisa, activist, UK	12/12/2013	Face-to-face (audio recorded)
Chris, unemployed/activist/ "digital miner up the North-West Passage," UK	1/27/2014	Face-to-face (audio recorded)

2. INTERVIEWS FOR FACEBOOK AUTOPOSTING APPS STUDY

TABLE A.2

Participant Information (Agreed Pseudonym, Age, Occupation, Country of Residence)	Interview Date	Format
Calum, duty manager, 30, UK	3/11/2014	Face-to-face (audio recorded)
Melanie, civil servant, 29, UK	3/17/2014	Face-to-face (audio recorded)
Sam, digital communications manager, 29, UK	4/30/2014	Face-to-face (audio recorded)
Marc, postgraduate student, 24, UK	5/20/2014	Face-to-face (audio recorded)
Beth, primary school teacher, 28, UK	5/19/2014	Face-to-face (audio recorded)
Sara, customer service manager, 30, UK	6/30/2014	Face-to-face (audio recorded)
Focus group one		
Alice, researcher, 28, UK	5/13/2014	Face-to-face (audio recorded)
Daniel, graphic designer, 29, UK	5/13/2014	Face-to-face (audio recorded)
Rory, sales manager, 30, UK	5/13/2014	Face-to-face (audio recorded)
Kevin, accounts executive, 25, UK	5/13/2014	Face-to-face (audio recorded)

Participant Information (Agreed Pseudonym, Age, Occupation, Country of Residence)	**Interview Date**	**Format**
Focus group two		
Sophie, publishing assistant, 28, UK	6/5/2014	Face-to-face (audio recorded)
Rebecca, lecturer in EAP, 27, UK	6/5/2014	Face-to-face (audio recorded)
Terry, graphic designer/carer, 28, UK	6/5/2014	Face-to-face (audio recorded)
Steve, trainee surveyor, 29, UK	6/5/2014	Face-to-face (audio recorded)
Audrey, marketer, 29, UK	6/5/2014	Face-to-face (audio recorded)
TP, producer, 29, UK	6/5/2014	

3. INTERVIEWS AND STUDY TIMELINE: GOOGLE APP STUDY

TABLE A.3

Participant Interview Organization	
Focus group participants	Tariq, 18, Dubai
(Agreed pseudonym, age, country of origin)	Rachel, 18, UK
	Heena, 18, Malaysia
	Lisa, 18, UK
One-on-one interview, participant 1	Giovanni, 18, Italy
One-on-one interview, participant 2	Laura, 18, UK

TABLE A.4

Study Timeline	
9/1/2014	Participants recruited to study and asked to activate the app/email screenshots, previous and allowed usage of app established.
10/8/2014	**Interview/focus group Session One**: Introductory questions/ discussion, semi-structured questions on app use.
10/15/2014	**Interview/focus group Session Two:** Semi-structured questions on app use, diary-writing exercise.
10/22/2014	**Interview/focus group Session Three:** Semi-structured questions on app use, Google Ad Settings exercise, experiment with Tariq.
10/29/2014	No interview session: students on Reading Week.
11/5/2014	**Interview/focus group Session Four:** Semi-structured questions on app use, discussion on personalization, privacy, and filter bubble theory (note: students attended lecture on privacy/ personalization the morning before the last session took place).

NOTES

CHAPTER 1

1. A cookie notice is a notification that alerts users to the presence of cookies—small text files used to collect anonymous, pseudonymous, and personal data from users—on the website they are visiting. The 2018 GDPR privacy law means that EU sites must alert users to the presence of cookies on their websites and gain consent from the user to use these cookies for data collection.
2. Based on the Alexa 500 list for most visited global sites (Alexa, 2018). The sites are Google.com (Google, 2017), YouTube.com (Google, 2017), Facebook.com (Facebook, 2017), Baidu.com (Baidu, 2019), Wikipedia.com, Yahoo.com (Yahoo, 2017), Qq.com (Qq.com, 2019), Taobao.com, Tmall.com (Tmall.com, 2018), and Twitter.com (Twitter, 2017). The only site that does not include the preceding terms in its privacy policy is Wikipedia, which is famously not-for-profit.
3. Jordan describes recursion as a process in which "information can eat itself [and] . . . in this way produce more information" (2015, 30); that is, recursion affords "the ability to take on digital information and then use it again and again to change similar digital actions" (2015, 31). Jordan explores some of these ramifications in relation to exploitation of information by data controllers; however, the reactive, feedback-able nature of recursion is also useful for considering how identities/algorithmic identities are co-constituted and co-related.
4. Barad (2007) and Haraway (1992) use the term "diffraction" rather than "reflection" in relation to methodological approaches to knowledge production. Barad argues that "for all of the recent emphasis on reflexivity as a critical method of self-positioning it remains caught up in geometries of sameness; by contrast, diffractions are attuned to differences—differences that our knowledge-making practices make and the effects they have on the world" (2007, 72). I believe that for the purposes of this book, both of these terminologies, despite their nuanced differences, sufficiently emphasize that the role of the researcher does not lie outside the meaning-making process.

CHAPTER 2

1. Data analytics team Kobell took the prize, improving the platform recommendations by 10%.
2. For example, watching *Orange Is the New Black* will result in being delivered suggestions in a genre category called "TV shows with strong female leads," with, for instance, comic crime drama *Jessica Jones* being inferred as an "87% match."
3. "Banner" and "side" ads are the ads commonly displayed on web pages in the margins above, below, and around the sides of a web page's central content.
4. The world wide web is a hyperlinked information system that can be used to access the internet.
5. In some ways, these discourses of individualism and personal freedom draw parallels with some of the non-commercial ideals already established by early net users—for example, Barlow's *Declaration for the Independence of Cyberspace* advocated that the web be a space that could "naturally" advocate individual freedom, free from state intervention (1997).
6. As I detail in Chapter 4, popular tracker blockers include Ghostery, LightBeam, and DoNotTrack, which all prevent the storage of some cookies on users' computers.
7. Web 2.0 was widely considered as the "next stage" of the world wide web. The term refers to "a combination of a) improved communication between people via social-networking technologies b) improved communication between separate software applications . . . and c) improved web interfaces that mimic the real-time responsiveness of desktop applications" (Roush, 2006). Though scholars such as Keen (2006) contest that the term is more of an ideology than a set of tangible technological developments, the dawn of web 2.0 was popularly seen as a departure from the text-based forms of communication commonly mobilized in web 1.0. Bassett notes that "2.0 is understood by Tom O'Reilly, who framed the term, as a technical and business corrective to the shortcomings of the early Internet" (2008, n.p.). Bassett considers O'Reilly's framing as "based on an understanding of the dynamics of the system (the new media ecology) in *use*" (2008, n.p.), and problematizes not only the industrial/practical affordances of web 2.0, but also its "cultural stakes" in order to consider the "participatory dynamics of the media system as a whole" (2008, n.p.).

CHAPTER 3

1. Foucault describes "technologies of the self" as those "which permit individuals to effect by their own means or with the help of others a certain number of operations on their own bodies and souls, thoughts, conduct and way of being, so as to transform themselves in order to attain a certain state of happiness, purity, wisdom, perfection or immortality" (1988, 18). It seems important to note that these technologies are not just tools, but also techniques—Foucault notes that reflection, penitence, and self-disclosure, as well as practices such diary writing, all constitute socio-historically specific operations of self-constitution/expression. Therefore the "technologies of the self" that Facebook offers are not just the tools (for uploading

photos, etc.) but also the operations for disclosure, articulation, and "self-knowing" (Foucault, 1988, 23) that these tools permit.

2. Lomborg and Bechmann define an API (Application Programming Interface) as "an interface of a computer program that allows the software to 'speak' with other software. This enables the development and enhancement of core social media services, for example, by allowing third-party companies to develop their own software clients for using Twitter or integrating Facebook with other social media service" (2014, 256).

CHAPTER 4

1. As a soon-to-be unemployed activist, campaigner, and non-university-affiliated intellectual, participant Chris was reluctant to give himself a job description—in the end we settled upon "activist," "unemployed" and "digital miner up the North-West Passage," though none of these descriptions fully encompasses Chris's multi-faceted occupational positions.
2. Ghostery's website tagline has changed a number of times since I first embarked on this research: its current pitch has dropped the tagline "Knowledge + Control = Privacy," in favor of a more straightforward offer to provide users with "[f]aster, safer, and smarter browsing" (Ghostery, 2018). However, at the time of the interviews, the privacy sum was not only foregrounded in the sites' marketing materials, its parts were also frequently referred to by the individuals interviewed for this study.
3. Ghostery is just one "self-help tool" available. Other notable tracker blockers include Tor, Adblocker Plus, Disconnect and LightBeam.
4. See Appendix 1 for full interview list and further interview details.
5. Eight of the twelve participants were recruited via Ghostery's retweet/reblog of my call. The other participants were recruited through "snowball sampling" (Browne, 2005), my own social network, and through digital arts/privacy events such as "Cryptoparties." I decided to use these extra recruitment methods as I felt that only interviewing individuals who responded to the blog call might mean that respondents would be essentially "fans" of Ghostery, whereas I wanted to try to capture the experiences of those who were engaged with Ghostery to a lesser extent (and thus did not follow their blog). This proved valuable—for example, Chris and Claire, who were both recruited outside of the blog, offered some interesting, perhaps more cynical views on the tracker blocker than some of the other participants.
6. At the time of interviews (late 2013 to early 2014), "knowledge" was featured as part of Ghostery's rhetorical sum and was a key term used to define the benefits of tracker blocker use. Since then the tracker and website have undergone a number of design changes, with "knowledge" enjoying less and less prominence in its informational material. However, the site continues to claim that the add-on allows users to "view" trackers, implying that insight, rather than knowledge of data tracking, is possible through the use of this blocker. In some ways, then, the site's development reflects the findings of this chapter: that though tracker blocking can afford users a

partial view to who is tracking them, it produces epistemic uncertainty, rather than increased knowledge as to how they are being tracked, managed, and anticipated.

7. The term has an interesting double meaning in that "power users" are treated as both those who need more "powerful" computers (Bhargarve and Feng, 2005) and those who are more "powerful" in the sense that they are more skilled than "average" users (Sundar and Marathe, 2010). Though I would argue that having more "skill" and "needing" more powerful computational tools should by no means equate to the idea that a user is more "powerful," it is nonetheless useful to use this term in the context of data tracking. This is because the "knowledge" that usually equates to "power" in the term "power user" is thoroughly problematized by the epistemological impossibility of "knowing" in any robust sense what happens to your data, as explored in the following.
8. For work on the uncanny and cyberspace, see Zylinska's theorization of cyberspace as "intrinsically uncanny" (2001, 161) in *On Spiders, Cyborgs and Being Scared*; Causey's analysis of uncanny performance in cyberspace (1999); or Vidler's work on virtual space as uncanny in relation to architecture (1992).
9. It is interesting to note that the term "privacy" seems to take a less dominant position on Ghostery's home page as the page has changed in the last three years. Again, the epistemic uncertainities bound in data tracking suggest that such a downplay may be connected to the fact that securing meaningful privacy is increasingly difficult under ubiquitous tracking practices.
10. Autofill text systems refer to the automatic generation of search engine suggestions (and other algorithmically inferred content) enacted in order to save the user time. For example, Google Search uses autofill to algorithmically "suggest" popular and personalized search queries to a particular user when they start typing a search query into the Google search bar. As Finn (2017) notes, the outputs of so-called personalized tools are in fact generated through collective data aggregation (autofill works by billions of collated search queries) and are often used to generate revenue as well as aid users.
11. McGeveran (2015) highlights that the EU and US have strikingly different privacy laws; in the US, privacy laws are based on the idea of individual freedom to act without encroachment from state or commercial interventions, while the EU bases its privacy laws on the premise that individuals have the right to protect and control information about themselves. McGeveran argues that this difference has led to legislation that treats privacy as a fundamental human right in the EU, whereas in the US privacy is geared more toward freedom of commercial interest.

CHAPTER 5

1. Facebook's international developer conference.
2. Sam preceded her use of the term with the question "Is it ridiculous if I say pretentious media studies words?" suggesting that her mobilization of this somewhat complex phrase can be explained by a background knowledge in theories of identity construction.

3. Perhaps Calum was right to be wary—as Beth explained in her interview, the only friend she has blocked on her News Feed was a friend who posted too much political content.
4. Jenkins's (2002) work on Bourdieu clearly demarks cultural capital as a matter of taste—he writes of Bourdieu's work *Distinction* that "Bourdieu's target here is . . . the consistent use of notions of 'taste' as a sort of naturally occurring phenomenon—to mark and maintain (in part by masking the marking) social boundaries, whether these be between the dominant and dominated classes or within classes" (2002, 135). In defining the accumulation of cultural capital as a matter of taste, it is possible to see how identity performance (by autopost or by Facebook users themselves) might be considered to be "tacky" or "tasteless."

CHAPTER 6

1. Nonetheless, some students may have had some exposure to critical media studies through their A levels or other means. However, the motivation behind interviewing them early in their university studies was not so much to catch them as "clean slates" in terms of their critical understanding of digital media—as Maynard (1994) notes, all research participants are situated within specific sociocultural contexts and so trying to capture "clean" experience would be a misguided methodological assumption. I hoped instead to chart their simultaneous development as new critical media scholars and as users of the Google app. I do believe, however, that trying the same methodological approach with second- or third-year students would have been less successful—their experience of university life and their developed attitude toward critical media studies may have prompted the interview participants to "perform" their expertise as scholars rather than express their engagement with the app as users.
2. Giovanni and Tariq both owned Android phones that come pre-installed with Google Now, while Laura, Heena, Lisa, and Rachel had iPhones, wherein they had to download the app to gain access. This may explain why the latter four participants had not used the app before.
3. The focus group/one-to-one session dynamic transpired partly as a methodological plan and partly to address timings and practicalities for the participants. Methodologically, the focus group vs. one-to-one dynamics enabled the capture of group dynamics and collective discussion about the app's predictions, while the one-to-one sessions were designed to allow for more intimacy and individual insight with individual subjects.
4. Location names have been changed to ensure participant anonymity.
5. It should be acknowledged here that although Rachel's self-recognition might be secured through Google Now, her feeling of the "lack" of Google Now's recognition might be produced through social interactions with other participants in the study. Heena, for example, received far more cards, despite making less effort to turn to face (Gillespie, 2014, 87) Google, and Rachel knows this. I would argue that though the lack of recognition may be the result of comparisons with other participants, it is interesting that her Google profile works to pleasurably secure her sense of self, rather than create privacy concerns, for example.

6. The fact that the participants did not feel that their studies equated to "work" is indicative of a wider societal assumption that students do not "work" in the same manner of paid labor. A wider critique of this assumption lies beyond the scope of this small study, yet it seems clears that the Google app is operating on the normative implication that users will have a "work" place that they "commute" to.

BIBLIOGRAPHY

Abbate J. (1999) *Inventing the Internet.* Cambridge, MA: MIT Press.

Abbate, J. (2010) "Privatizing the Internet: Competing Visions and Chaotic Events, 1987–1995." *IEEE Annuals of the History of Computing* 32: 10–22.

Adorno, T. ([1952] 1994) *"The Stars Down to Earth" and Other Essays on the Irrational in Culture.* London: Routledge.

Alexa. (2018) *Top Global Websites.* Available at: http://www.alexa.com/topsites (accessed September 7, 2018).

Agre, P. (1994). "Surveillance and capture: Two models of privacy". *The Information Society*, 10(2): 101–127.

Amazon. (2017) *Alexa Terms of Use.* Available at: https://www.amazon.co.uk/gp/help/customer/display.html?nodeId=201809740 (accessed July 3, 2017).

Amazon. (2018) "What Is Echo Look?" *Amazon.* Available at: https://www.amazon.com/Amazon-Echo-Look-Camera-Style-Assistant/dp/B0186JAEWK (accessed September 2, 2018).

Andrejevic, M. (2011) "Social Network Exploitation." In Z. Papacharissi (ed.), *A Networked Self.* New York: Routledge, pp. 82–102.

Andrejevic, M. (2013) *Infoglut: How Too Much Information Is Changing the Way We Think and Know.* New York: Routledge.

Android. (2012) "Introducing Google Now." *YouTube.* Available at: https://www.youtube.com/watch?v=pPqliPzHYyc (accessed September 7, 2018).

AOL. (2017) "UK Privacy Policy." *AOL.* Available at: http://privacy.aol.co.uk/ (accessed April 4, 2017).

Arora, N., Dreze, X., Ghose, A., Hess, J. D., Iyengar, R., Jing, B., & Zhang, Z. J. (2008) "Putting One-to-One Marketing to Work: Personalization, Customization, and Choice." *Marketing Letters* 19(3–4): 305–321. http://doi.org/10.1007/s11002-008-9056-z

AstraZeneca. (2015) *Home.* Available at: http://www.astrazeneca.co.uk/home (accessed April 26, 2015).

Baidu (2019) *Privacy Policy*. Available at: http://usa.baidu.com/privacy/ (accessed November 5, 2019).

Barad, K. (2007) *Meeting the Universe Halfway: Quantum Physics and the Entanglement of Matter and Meaning*. Durham, NC: Duke University Press.

Barlow, J. P. (1996) *A Cyberspace Independence Declaration*. Available at: http://w2.eff.org/Censorship/Internet_censorship_bills/barlow_0296.declaration (accessed August 27, 2014).

Bassett, C. (1997) "Virtually Gendered: Life in an Online World." In S. Thornton & Gelder (eds.), *The Subcultures Reader*, 537–500. London: Routledge.

Bassett, C. (2007) *The Arc and the Machine*. Manchester: Manchester University Press.

Bassett, C. (2008) "FCJ-088 New Maps for Old? The Cultural Stakes of '2.0.'" *Fibreculture Journal* 13: After Convergence.

Bassett, C. (2013) "Silence, Delirium, Lies?" *First Monday* 18(3). http://doi.org/10.5210/fm.v18i3.4617

Bassett, C., Fotopoulou, A., & Howland, K. (2015) "Expertise: A Report and a Manifesto." *Convergence* 21(3): 328–342.

Baumer, M., Ames, G., Brubaker, J. R., Burrell, J., & Dourish, P. (2015) "Why Study Non-Use?" *First Monday* 20: 11. doi: http://dx.doi.org/10.5210/fm.v20i11.6310.

BBC. (2014) "Facebook Wants to 'Listen' to Your Music and TV." *BBC News*, May 22 [Online]. Available at: http://www.bbc.co.uk/news/technology-27517817 (accessed May 22, 2014).

BBC. (2014a) "Google to Face UK Users in Privacy Row." *BBC News*, January 16 [Online]. Available at: http://www.bbc.co.uk/news/technology-25763000 (accessed October 2, 2016).

BBC. (2015) "Personalised BBC News App on Your Phone & Tablet." *BBC News*, June 29 [Online]. Available at: http://www.bbc.co.uk/news/help-33449119 (accessed October 2, 2016).

BBC. (2016) "Edward Snowden: Timeline." *BBC News*, August 20 [Online]. Available at: http://www.bbc.co.uk/news/world-us-canada-23768248 (accessed April 26, 2016).

Beck, U. (1992) *Risk Society: Towards a New Modernity*. London: Sage.

Benkler, Y. (2006) *The Wealth of Networks: How Social Production Transforms Markets and Freedom*. New Haven, CT: Yale University Press.

Berry, D. (2012) "The Social Epistemologies of Software." *Social Epistemology* 26(3–4): 379–398.

Berry, D. (2014) *Critical Theory and the Digital*. London: Bloomsbury.

Best, K., & Tozer, N. (2012) "Scaling Digital Walls: Everyday Practices of Consent and Adaptation to Digital Architectural Control." *International Journal of Cultural Studies*, 401–417. doi:10.1177/1367877912460618.

Bettig, R. V. (1996) *Copyrighting Culture: The Political Economy of Intellectual Property*. Boulder, CO: Westview Press.

Bhargava, H. K., & Feng, J. (2005) "America OnLine's Internet Access Service: How to Deter Unwanted Customers." *Electronic Commerce Research and Applications* 4(1): 35–48. http://doi.org/10.1016/j.elerap.2004.10.008.

Black, I. (2013) "NSA Spying Scandal: What We Have Learned." *The Guardian*, June 10 [Online]. Available at: http://www.theguardian.com/world/2013/jun/10/nsa-spying-scandal-what-we-have-learned (accessed September 2013).

Blom, J. (2000) "Personalisation: A Taxonomy." *Computer Human Interaction* 2000 Conference, York.

Blondel, J. (2010) *Political Leadership, Parties and Citizens: The Personalisation of Leadership*. London: Routledge.

Bodle, R. (2014) "Predictive Algorithms and Personalization Services on Social Network Sites: Implications for Users and Society." In A. Bechmann & S. Lomborg (eds.), *The Ubiquitous Internet: User and Industry Perspectives*, 130–145. New York: Routledge.

Boellstorff, T. (2008) *Coming of Age in Second Life: An Anthropologist Explores the Virtually Human.* Princeton, NJ: Princeton University Press.

Boler, M. (2007) "Hypes, Hopes and Actualities: New Digital Cartesianism and Bodies in Cyberspace." *New Media and Society* 9(1): 139–168.

Bolin, G., & Andersson Schwarz, J. (2015) "Heuristics of the Algorithm: Big Data, User Interpretation and Institutional Translation." *Big Data and Society* 2. doi:10.1177/2053951715608406.

Bourdieu, P. ([1979] 1989) *Distinction: A Social Critique of the Judgement of Taste.* London: Routledge.

Bourdieu, P. ([1984] 1998) "Distinction and the Aristocracy of Culture." In J. Storey (ed.), *Cultural Theory and Popular Culture: A Reader,* 125–153. Hemel Hempstead: Harvester Wheatsheaf.

Boyd, D. (2014) *It's Complicated: The Social Lives of Networked Teens*. New Haven: Yale University Press.

Boyd, D. & Crawford, K. (2012) "Critical Questions for Big Data". *Information, Communication & Society*, 15(5): 662–679. doi: 10.1080/1369118X.2012.678878

Brand, S. (1988) *The Media Lab: Inventing the Future at M.I.T.* (reprint edition). New York: Penguin Books.

Braun, V., & Clarke, V. (2013) *Successful Qualitative Research: A Practical Guide for Beginners.* London: Sage.

Browne, K. (2005) "Snowball Sampling: Using Social Networks to Research Non-Heterosexual Women." *International Journal Social Research Methodology* 8: 47–60. doi:10.1080/1364557032000081663.

Brunton, F., & Coleman, G. (2014) "Closer to the Metal." In G. Tarleton, B. Boczkowski, & K. Foot (eds.), *Media Technologies: Essays on Communication, Materiality, and Society*, 77–98. Cambridge, MA: MIT Press.

Brunton, F., & Nissenbaum, H. (2011) "Vernacular Resistance to Data Collection and Analysis: A Political Theory of Obfuscation." *First Monday* 16.

Brunton, F., & Nissenbaum, H. (2015) *Obfuscation: A User's Guide to Privacy and Protest.* Cambridge, MA: MIT Press.

Bucher, T. (2016) "The Algorithmic Imaginary: Exploring the Ordinary Affects of Facebook Algorithms." *Information, Communication & Society* 20(1): 30–44. doi:10.1080/1369118X.2016.1154086.

Burkhalter, B. (1999) "Reading Race Online: Discovering Racial Identity in Usenet Discussions." In A. Marc & P. Kollock (eds.), *Communities in Cyberspace*. London: Routledge, 60–75.

Burns, R. (2015) "Tablets: Specific, Generic, Perfect?" In C. Bassett, K. O'Riordan, K. Grant, & R. Burns (eds.), *The Tablet Book*, 206–231. Sussex: REFRAME Publications.

Business Insider. (2014) "Mark Zuckerberg Wants to Build a "Perfect Personalized Newspaper." *Business Insider.* Available at: http://uk.businessinsider.com/mark-zuckerberg-wants-to-build-a-perfect-personalized-newspaper-2014-11?r=US&IR=T (accessed February 11, 2015).

Butler, J. (1988) "Performative Acts and Gender Constitution: An Essay in Phenomenology and Feminist Theory." *Theatre Journal* 40(4): 519–531.

Butler, J. (1989) "Foucault and the Paradox of Bodily Inscriptions." *The Journal of Philosophy* 86(11): 601–607.

Butler, J. (1990) *Gender Trouble*. London: Routledge.

Butler, J. (1993) *Bodies That Matter: On the Discursive Limits of "Sex."* London: Routledge.

Butler, J. (1998) "How Bodies Come to Matter: An Interview with Judith Butler." *Signs: Journal of Women in Culture and Society* 23(2): 275.

BuzzFeed. (2018) "BuzzFeed's Privacy Notice and Cookie Policy,." *BuzzFeed*. Available at: https://www.buzzfeed.com/about/privacy (accessed September 2, 2018).

Campanelli, V. (2014) "Frictionless Sharing: The Rise of Automatic Criticism." In R. Konig & M. Rasch. (eds.), *Society of the Query Reader: Reflections on Web Search*, 41–48. Amsterdam: Institute of Network Cultures.

Campbell-Kelly, M., & Garcia-Swartz, D. D. (2013) "The History of the Internet: The Missing Narratives." *Journal of Information Technology* 28(1): 18–33.

Cashmore, P. (2009) "Privacy Is Dead, and Social Media Hold Smoking Gun." *CNN*. Available at: http://edition.cnn.com/2009/OPINION/10/28/cashmore.online.privacy/index.html (accessed September 4, 2018).

Causey, M. (1999) "The Screen Test of the Double: The Uncanny Performer in the Space of Technology." *Theatre Journal* 51(4): 383–394.

Cesca, B. (2013) "How *The Guardian* Is Quietly and Repeatedly Spying on You." *Huffington Post*, September 13 [Online] http://www.huffingtonpost.com/bob-cesca/how-the-guardian-is-quiet_b_3923408.html (accessed September 13, 2013).

Chan, A. (2015) "Big Data Interfaces and the Problem of Inclusion." *Media Culture & Society* 37: 1078–1083. doi:10.1177/0163443715594106.

Chen, L., Chung D., Chia, J., & Tseng, J. (2013) "The Effect of Recommendation Systems on Internet-Based Learning for Different Learners: A Data Mining Analysis." *British Journal of Educational Technology* 44(5): 758–773.

Cheney-Lippold, J. (2011) "A New Algorithmic Identity: Soft Biopolitics and the Modulation of Control." *Theory, Culture and Society* 28: 164–181.

Cheney-Lippold, J. (2017) *We Are Data: Algorithms and the Making of Our Digital Selves*. New York: New York University Press.

Christl, W., & Spiekermann, S. (2016) *Networks of Control: A Report on Corporate Surveillance, Digital Tracking, Big Data and Privacy*. Vienna: Facultas.

Chun, W. H. K. (2008) "On 'Sourcery,' or Code as Fetish." *Configurations* 16(3): 299–324.

Chun, W. H. K. (2015) *Updating to Remain the Same: Habitual New Media.* Cambridge, MA: MIT Press.

Cohen, N. (2013) "Commodifying Free Labor Online: Social Media, Audiences, and Advertising." In M. McAllister & E. West (eds.), *The Routledge Companion to Advertising and Promotional Culture.* Abingdon: Routledge, pp. 177–191.

Cohn, J. (2019) *The Burden of Choice: Recommendations, Subversion, and Algorithmic Culture.* New Brunswick, NJ: Rutgers University Press.

Cotterill, P. (1992) "Interviewing Women: Issues of Friendship, Vulnerability, and Power." *Women's Studies International Forum* 15(5–6): 593–606. http://doi.org/10.1016/0277-5395(92)90061-Y.

Cover, R. (2012) "Performing and Undoing Identity Online: Social Networking, Identity Theories and the Incompatibility of Online Profiles and Friendship Regimes." *Convergence* 18(2): 177–193.

Criteo. (2018) "Product Recommendations." *Criteo.* Available at: https://www.criteo.com/technology/criteo-engine/product-recommendations/ (accessed September 6, 2018).

Curran, J., & Seaton, J. (2010) *Power without Responsibility: Press, Broadcasting and the Internet in Britain: Press and Broadcasting in Britain.* London: Routledge.

Dean, J. (2005) "Communicative Capitalism: Circulation and the Foreclosure of Politics." *Cultural Politics* 1(1): 51–78.

De Certeau, M. ([1984] 2002) *The Practice of Everyday Life.* Berkeley: University of California Press.

Deighton, J., & Johnson, P. (2013) *The Value of Data: Consequences for Insight, Innovation & Efficiency in the U.S. Economy.* Report, Alexandria: ANA.

Deleuze, G. (1992) "Postscript on the Societies of Control." *October* 59: 3–7.

Deuze, M. (2011) "Media Life." *Media, Culture & Society* 33(1): 137–148.

Dim, E., Kuflik, T., & Reinhartz-Berger, I. (2015) "When User Modeling Intersects Software Engineering: The Info-Bead User Modeling Approach." *User Modeling and User-Adapted Interaction* 25(3): 189–229. http://doi.org/10.1007/s11257-015-9159-1.

Dixon, P., & Gellman, R. (2014): *The Scoring of America: How Secret Consumer Scores Threaten Your Privacy and Your Future.* World Privacy Forum, April 2, 2014. Available at: https://www.worldprivacyforum.org/2014/04/wpf-report-the-scoring-of-america-how-secret-consumer-scores-threaten-your-privacy-and-your-future/

Donath, J. S. (1999) "Identity and Deception in the Virtual Community." In A. Marc & P. Kollock (eds.), *Communities in Cyberspace.* London: Routledge, pp. 29–59.

Einstein, M. (2017) *Advertising: What Everyone Needs to Know.* Oxford: Oxford University Press.

Ekbia, H. R. (2016) "Digital Inclusion and Social Exclusion: The Political Economy of Value in a Networked World." *The Information Society* 32(3): 165–175. http://doi.org/10.1080/01972243.2016.1153009.

Electronic Arts. (2017) "Privacy Policy." *EA.* Available at: http://www.ea.com/privacy-policy (accessed June 23, 2017).

Electronic Frontier Foundation. (2013) *Defending Your Rights in a Digital World.* Available at: https://www.eff.org/ (accessed June 14, 2013).

Ellison, N., Steinfield, C., & Lampe, C. (2007) "The Benefits of Facebook 'Friends': Social Capital and College Students' Use of Online Social Network Sites." *Journal of Computer-Mediated Communication* 12: 1143–1168.

Emmel, N. (2013) "Sample Size." *Sampling and Choosing Cases in Qualitative Research: A Realist Approach*. Los Angeles: Sage.

Experian. (2018) "The Best Data Unlocks the Best Marketing." *Experian*. Available at: http://www.experian.com/marketing-services/targeting/data-driven-marketing.html (accessed September 6, 2018).

Facebook. (2017) *Privacy Policy*. Available at: https://www.facebook.com/legal/FB_Work_Privacy (accessed June 28, 2017).

Facebook. (2018) "Your Ad Preferences." *Facebook*. Available at: https://www.facebook.com/ads/preferences/?entry_product=ad_settings_screen (accessed September 2, 2018).

Facebook Apps Center. (2014) *Get Spotify*. Available at: https://apps.facebook.com/get-spotify/ (accessed February 14, 2014).

Facebook Newsroom (2014) *News Feed FYI: Giving People More Control over When They Share from Apps*. http://newsroom.fb.com/news/2014/05/news-feed-fyi-giving-people-more-control-over-when-they-share-from-apps/ (accessed August 15, 2014).

Facebook Newsroom (2014a) "Introducing Anonymous Login and an Updated Facebook Login." *Facebook Newsroom*. Available at: https://newsroom.fb.com/news/2014/04/f8-introducing-anonymous-login-and-an-updated-facebook-login/ (accessed November 1, 2019).

Facebook Developers (2018) "Facebook Day 1 Keynote". *YouTube*. Available at: https://www.youtube.com/watch?v=ldtuSYqgPLQ&list=PLb0IAmt7-GS0RLjX7uCpbqGOhrKYTJW_8 (accessed November 1, 2019).

Facebook Help Center. (2018) "What Kinds of Posts Will I See in News Feed?" *Help Center*. Available at: https://www.facebook.com/help/166738576721085?sr=20&query=NEWS%20FEED&sid=1r3FTDuXU4VAfCAmJ (accessed September 2, 2018).

Facebook Timeline. (2014) *Introducing Timeline*. Available at: https://www.facebook.com/about/timeline (accessed June 13, 2014).

Faden, R. R., & Beauchamp, T. L. (1986) *A History and Theory of Informed Consent*. Oxford: Oxford University Press.

Falahrastegar, M., Haddadi, H., Uhlig, S., & Mortier, R. (2014) "Anatomy of the Third-Party Web Tracking Ecosystem." Available at: http://arxiv.org/pdf/1409.1066v1.pdf (accessed August 23, 2014).

Fan, H., & Poole, M. (2006) "What Is Personalization? Perspectives on the Design and Implementation of Personalization in Information Systems." *Journal of Organizational Computing and Electronic Commerce* 16(3–4): 179–202.

Finn, E. (2017) *What Algorithms Want: Imagination in the Age of Computing*. Cambridge, MA: MIT Press.

Foucault, M. (1988) "Technologies of the Self." In *Essential Works of Foucault 1954–1984: Ethics*, 233–252. London: Penguin.

Foucault, M. (1988a) *The Care of the Self: The History of Sexuality*, Volume 3. London: Penguin.

Foucault, M. ([1974] 1995) *Discipline and Punish: The Birth of the Prison*. London: Routledge

Fuchs, C. (2010) "Labor in Informational Capitalism and on the Internet." *The Information Society* 26: 179–196.

Fuchs, C. (2012) "The Political Economy of Privacy on Facebook." *Television and New Media* 13(2): 139–159.

Fuez, M., Fuller, M., & Stalder, F. (2011) "Personal Web Searching in the Age of Semantic Capitalism: Diagnosing the Mechanics of Personalisation." *First Monday* 16(2).

Gates, B. (1995) *The Road Ahead*. New York: Viking.

Geertz, C. ([1973] 1993) "Thick Description: Towards an Interpretive Theory of Culture." In *The Interpretation of Cultures*, 310–323. London: Fontana Press.

Gerlitz, C., & Helmond, A., (2013) "The Like Economy: Social Buttons and the Data-Intensive Web." *New Media and Society*, 15(8), 1348–1365. doi:10.1177/1461444812472322.

Getting Personal. (2017) *Homepage*. Available at: http://www.gettingpersonal.co.uk/gifts?gclid=CjwKEAjw5J6sBRDp3ty_17KZyWsSJABgp-Oar-s3j_5g5ilsfpKMeamLBDP_Iu19ZejMuf2CzCxULhoCwK3w_wcB (accessed June 23, 2017).

Ghoshal, D. (2018) "Mapped: The Breathtaking Global Reach of Cambridge Analytica's Parent Company." *Quartz*. Available at: https://qz.com/1239762/cambridge-analytica-scandal-all-the-countries-where-scl-elections-claims-to-have-work.ed/ (accessed 12 December, 2018)

Ghostery. (2014) *Knowledge + Control = Privacy*. Available at: https://www.ghostery.com/en/ (accessed October 2013–February 2014).

Ghostery. (2018) "Faster, Safer, and Smarter Browsing." *Ghostery*. Available at: https://www.ghostery.com/ (accessed September 6, 2018).

Gibbs, S. (2015) "Facebook Tracks All Users, Breaching EU Law." *The Guardian*, March 31 [Online]. Available at: http://www.theguardian.com/technology/2015/mar/31/facebook-tracks-all-visitors-breaching-eu-law-report (accessed March 31, 2015).

Gibbs, S. (2016) "How Much Are You Worth to Facebook"? *The Guardian*, January 28 [Online]. Available at: http://www.theguardian.com/technology/2016/jan/28/how-much-are-you-worth-to-facebook (accessed January 28, 2016).

Gibson, J. J. (1986) *The Ecological Approach to Visual Perception*. Hillsdale, NJ: Lawrence Erlbaum Associates.

Giddens, A. (1991) *Modernity and Self-Identity: Self and Society in the Late Modern Age*. Stanford, CA: Stanford University Press.

Gillani, N., & Eynon, R. (2014) "Communication Patterns in Massively Open Online Courses." *The Internet and Higher Education* 23: 18–26.

Gillespie, T. (2014) "The Relevance of Algorithms." In G. Tarleton, B. Boczkowski, & K. Foot (eds.), *Media Technologies: Essays on Communication, Materiality, and Society*, 167–194. Cambridge, MA: MIT Press.

Gillmor, D. (2014) "As We Sweat Surveillance, Companies like Google Track Our Data." *The Guardian*, April 18 [Online]. Available at: http://www.theguardian.com/commentisfree/2014/apr/18/corporations-google-should-not-sell-customer-data (accessed May 2, 2014).

Gitelman, L. (2006) *Always Already New: Media, History, and the Data of Culture.* Cambridge, MA: MIT Press.

Goffman, I. (1959) *The Presentation of Self in Everyday Life*. London: Penguin.

Goffey, A. (2008) "Algorithm," In R. F. Malina & S. Cubitt (eds.), *Software Studies: A Lexicon*. Cambridge, MA: MIT Press, pp. 15–20.

Google. (2014) *Google Now.* Available at: https://www.google.com/landing/now/ (accessed September 7, 2018).

Google. (2015) *Usage and Diagnostics for Location History.* Available at: https://support.google.com/accounts/answer/3118687?ref_topic=3100928&hl=en (accessed May 7, 2015).

Google. (2016) "Algorithms: How Search Works." *Google.* Available at: https://www.google.co.uk/insidesearch/howsearchworks/algorithms.html (accessed April 20, 2016).

Google. (2017) "Welcome to the Google Privacy Policy." *Google.* Available at: https://www.google.co.uk/intl/en/policies/privacy/?fg=1 (accessed June 28, 2017).

Google Ad Settings. (2018) "How Your Ads Are Personalized." *Google Ad Settings.* Available at: https://adssettings.google.com/authenticated (accessed September 7, 2018).

Google. (2018a) "Google." *Google Play Store.* Available at: https://play.google.com/store/apps/details?hl=en&id=com.google.android.googlequicksearchbox (accessed September 7, 2018)

Google Home. (2018) "Overview." *Google Home.* Available at: https://store.google.com/product/google_home (accessed September 7, 2018).

Google Now Developer Schema. (2014) *Developer Schema.* Available at: https://developers.google.com/schemas/now/cards/ (accessed December 12, 2014).

Gravity. (2013) "The Personalized Web Powered by the Interest Graph." Available at: http://www.gravity.com/ (accessed May 5, 2015).

Gravity Insights. (2013) *Homepage.* Available at: http://www.gravity-insight.com/ (accessed December 12, 2013).

Greenfield, P. (2018) "The Cambridge Analytica Files: The Story So Far." *The Guardian*, March 25 [Online]. Available at: https://www.theguardian.com/news/2018/mar/26/the-cambridge-analytica-files-the-story-so-far (accessed September 4, 2018).

Greenstein, R., & Esterhuysen, A. (2006) "The Right to Development in the Information Society." In R. K. Jørgensen (ed.), *Human Rights in the Global Information Society.* Cambridge, MA: MIT Press, pp. 281–302.

Grosser, B. (2014) "How the Technological Design of Facebook Homogenizes Identity and Limits Personal Representation." *Hz-Journal* 19. Available at: http://www.hz-journal.org/n19/grosser.html.

Guardian. (2013) "Edward Snowden." *The Guardian.* Available at: http://www.theguardian.com/us-news/edward-snowden (accessed June 14, 2013).

Guardian. (2018) "What Should I Do about All the GDPR Pop-Ups on Websites?" *The Guardian*, July 5 [Online]. Available at: https://www.theguardian.com/technology/askjack/2018/jul/05/what-should-i-do-about-all-the-gdpr-pop-ups-on-websites (accessed May 11, 2019)

Gubrium, J. F., & Holstein, J. A. (2001) *Handbook of Interview Research*. Thousand Oaks, CA: SAGE.

Hacking, I. (1986) "Making Up People." In T. C. Heller & C. Brooke-Rose (eds.), *Reconstructing Individualism: Autonomy, Individuality, and the Self in Western Thought*. Stanford, CA: Stanford University Press.

Hagel, J., & Armstrong A. (1997) *Net Gain: Expanding Markets through Virtual Communities*. Boston: Harvard Business School Press.

Hall, S. ([1980] 2009) "Encoding/ Decoding." In S. Thornham, C. Bassett, & P. Marris (eds.), *Media Studies: A Reader*. New York: New York University Press.

Hall, S. ([1989] 1996) "The Problem of Ideology: Marxism without Guarantees." In D. Morley & K. Chen (eds.), *Stuart Hall: Critical Dialogues in Cultural Studies*. London: Routledge.

Hall, S. (1996) "What Is Identity?" In S. Hall & P. Gay (eds.), *Questions of Cultural Identity*. London: SAGE.

Hallinan, B., & Striphas, T. (2016) "Recommended for You: The Netflix Prize and the Production of Algorithmic Culture." *New Media & Society* 18(1): 117–137. http://doi.org/10.1177/1461444814538646.

Haraway, D. (1988) "Situated Knowledges: The Science Question in Feminism and the Privilege of Partial Perspective." *Feminist Studies* 14(3): 575–599.

Haraway, D. (1991) *Simians, Cyborgs and Women: The Reinvention of Nature*. New York: Free Association Books.

Hardt, M., & Negri, A. (2012) *Declaration*. New York: Argo.

Hardy, J. (2013) "The Changing Relationship between Media and Marketing." In H. Powell (ed.), *Promotional Culture and Convergence: Markets, Methods, Media*. London: Routledge, pp. 125–150.

Hardy, J. (2015) "Critical Political Economic Approaches to Advertising." In C. Wharton (ed.), *Critical Advertising*. London: Routledge, pp. 65–84.

Hartley, D. (2012) *Education and the Culture of Consumption: Personalisation and the Social Order*. London: Routledge.

Hamburger, E. (2014) "Facebook Gives Up Sharing Everything You Do Online." *The Verge*, May 27 [Online]. Available at: http://www.theverge.com/2014/5/27/5754862/facebook-gives-up-on-automatically-sharing-everything-you-do-online-open-graph (accessed August 3, 2014).

Hearn, A. (2017) "Verified: Self-Presentation, Identity Management, and Selfhood in the Age of Big Data." *Popular Communication: The International Journal of Media and Culture* 15(2), 62–77. doi: https://doi.org/10.1080/15405702.2016.1269909.

Hern, A., & Mahdawi, A. (2018) "Beware the Smart Toaster: 18 Tips for Surviving the Surveillance Age." *The Guardian*, March 28 [Online]. Available at: https://www.theguardian.com/technology/2018/mar/28/beware-the-smart-toaster-18-tips-for-surviving-the-surveillance-age (accessed September 10, 2017).

Hernæs, C (2017) "What Facebook's European payment license could mean for banks." *TechCrunch*. January 12 [Online]. Available at: https://techcrunch.com/2017/01/12/what-facebooks-european-payment-license-could-mean-for-banks/ (accessed 15 June, 2018).

Highmore, B. (2012) "Everyday Life as a Critical Concept in Everyday Life." *Critical Concepts in Media and Cultural Studies* 1: 1–16.

Hillis, K., Petit, M., & Jarrett, K. (2013) *Google and the Culture of Search*. London: Routledge.

Hine, C. (2000) *Virtual Ethnography*. London: SAGE.

Hine, C. (2015) *Ethnography for the Internet: Embedded, Embodied and Everyday.* London: Bloomsbury.

Horling, B., & Kulick, M. (2009) "Personalized Search for Everyone." *Google Blog.* Available at: https://googleblog.blogspot.co.uk/2009/12/personalized-search-for-everyone.html (accessed September 2, 2018).

Hosanagar, K., Fleder, D., Lee, D., & Buja, A. (2013) "Will the Global Village Fracture into Tribes? Recommender Systems and Their Effects on Consumer Fragmentation." *Management Science* 60: 805–823.

Houtman, P. D., Koster, M. W., & de Aupers, P. S. (2013) *Paradoxes of Individualization: Social Control and Social Conflict in Contemporary Modernity.* London: Ashgate.

Huffington Post. (2015) *HuffingtonPost.co.uk Privacy Policy.* Available at: http://www.huffingtonpost.com/p/huffingtonpostcouk-privacy-policy.html (accessed May 5, 2015).

iGoogle. (2013) *iGoogle Personalised Homepage*. Available at: https://developers.google.com/igoogle/?hl=en (accessed June 10, 2013).

Ignatow, G., & Robinson, L. (2017) "Pierre Bourdieu: Theorizing the Digital." *Information, Communication & Society* 20(7): 950–966. doi:10.1080/1369118X.2017.1301519.

Israel, M., & Hay, I. (2006) *Research Ethics for Social Scientists.* London: SAGE.

Jannach, D., Ludewig, M., & Lerche, L. (2017) "Session-Based Item Recommendation in e-Commerce: On Short-Term Intents, Reminders, Trends and Discounts." *User Modeling and User-Adapted Interaction* 27(3–5): 351–392.

Jarrett, K. (2014) "A Database of Intention?" In R. Konig & M. Rasch (eds.), *Society of the Query Reader: Reflections on Web Search*, 16–29 Amsterdam: Institute of Network Cultures.

Jarrett, K. (2014a) "The Relevance of 'Women's Work': Social Reproduction and Immaterial Labour in Digital Media." *Television and New Media* 15(1): 14–29.

Jenkins, H. (2006) *Convergence Culture: Where Old and New Media Collide.* New York: New York University Press.

Jenkins, R. (2002) *Pierre Bourdieu.* London: Routledge.

Jones, O. (2011) *Chavs: The Demonization of the Working Class.* London: Verso.

Jordan, T. (2013) *Internet, Society and Culture: Communicative Practices before and after the Internet.* London: Bloomsbury.

Jordan, T. (2015) *Information Politics: Liberation and Exploitation in the Digital Society.* London: Pluto Press.

Kang, H., & Shin, W. (2016) "Do Smartphone Power Users Protect Mobile Privacy Better than Nonpower Users? Exploring Power Usage as a Factor in Mobile Privacy Protection and Disclosure". *Cyberpsychology, Behavior and Social Networking,* 19(3): 179–185.

Kant, T. (2014) "Giving the 'Viewser' a Voice? Situating the Individual in Relation to Personalization, Narrowcasting, and Public Service Broadcasting." *Journal of Broadcasting & Electronic Media* 58(3): 381–399. doi:10.1080/08838151.2014.935851.

Karppi, T. (2013) "FCJ-166 'Change Name to No One. Like People's Status': Facebook Trolling and Managing Online Personas." *The Fibreculture Journal* 22: 278–300

Keen, A. (2006) "Web 2.0". *Weekly Standard*. Available at: http://www.weeklystandard.com/Content/Public/Articles/000/000/006/714fjczq.asp?pg=2 (accessed November 5, 2019).

Kelion, L. (2018) "How to Handle the Flood of GDPR Privacy Updates." *BBC News*, April 28 [Online]. Available at: https://www.bbc.co.uk/news/technology-43907689 (accessed September 2, 2018).

Kellner, D. M., & Durham, M. G. (2001) "Adventures in Media and Cultural Studies: Introducing the Keyworks." In M. G. Durham & D. M. Kellner (eds.), *Media and Cultural Studies Keyworks*. London: Blackwell, pp. ix–xxxviii.

Kennedy, H. (1999) "Beyond Anonymity, or Future Direction for Internet Identity Research." In S. Thornham, C. Bassett, & P. Marris (eds.), *Media Studies: A Reader*. Edinburgh: Edinburgh University Press, pp. 825–838.

Kennedy, J., Nansen, B., Arnold, M., Wilken, R., & Gibbs, M. (2015) "Digital Housekeepers and Domestic Expertise in the Networked Home." *Convergence: International Journal of Research in New Media Technologies*, 21(4): 408–422. doi:10.1177/1354856515579848

Killoran, J. B. (2002) "Under constriction: Colonization and synthetic institutionalization of Web space". *Computers and Composition*, 19: 19–37.

Kiskis, M. (2018) "GDPR Is Eroding Our Privacy, Not Protecting It." *The Next Web*. Available at: https://thenextweb.com/contributors/2018/08/05/gdpr-privacy-eroding-bad/ (accessed September 10, 2018).

Kim, H., & Saddik A. (2013) "Exploring Social Tagging for Personalized Community Recommendations." *User Modelling and User-Adaptive Interactions* 23: 249–285. doi:10.1007/s11257-012-9130-3.

King, R. (2014) "Facebook dials back on third-party app shares." *ZDNet*. Available at: https://www.zdnet.com/article/facebook-dials-back-on-third-party-app-shares/ (accessed: November 1, 2019).

Kitchin, R., & Dodge, M. (2011) *Code/Space: Software in Everyday Life*. Cambridge, MA: MIT Press.

König, P. D. (2017) "The Place of Conditionality and Individual Responsibility in a 'Data-Driven Economy.'" *Big Data & Society*, 4(2): 1–4.

Koutra, D., Bennet, P. & Horvitz, E. (2014). "Events and Controversies: Influences of a Shocking News Event on Information Seeking." *EEES* 11. http://arxiv.org/pdf/1405.1486v1.pdf.

Kvale, S., & Brinkmann, S. (2009) *InterViews*. Thousand Oaks, CA: SAGE.

Lapenta, G & Jørgensen, R. (2015) "Youth, privacy and online media: Framing the right to privacy in public policy-making". *First Monday*, (20)3. doi:https://doi.org/10.5210/fm.v20i3.5568.

Lanier, J. (2010) *You Are Not a Gadget: A Manifesto*. New York: Penguin.

Latour, B., as Johnson, J. (1988) "Mixing Humans and Nonhumans Together: The Sociology of the Door Closer." *Social Problems* 35(3): 298–310.

Latour, B. (1999) *Pandora's Hope: Essays on the Reality of Science Studies*. Cambridge, MA: Harvard University Press.

Latour, B. (2005) *Assembling the Social: An Introduction to Actor Network Theory.* Oxford: Oxford University Press.

Law, J. (1991) *A Sociology of Monsters: Essays on Power, Technology and Domination.* London: Routledge.

Levy, S. (2011) *In The Plex: How Google Thinks, Works, and Shapes Our Lives.* New York: Simon & Schuster.

Lightbeam. (2013) "Lightbeam." http://www.mozilla.org/en-US/lightbeam/ (accessed December 21, 2013).

Liu, H. (2008) "Social Network Profiles as Taste Performances." *Journal of Computer-Mediated Communication* 13: 252–275.

Live.com. (2017) *Live Homepage* (Cookie Notice). Available at: https://login.live.com. (accessed December 10, 2017).

Livingstone, S. (2014) "Identifying the Interests of Digital Users as Audiences, Consumers, Workers and Publics." In T. Gillespie, P. Boczkowski, & K. Foot (eds.), *Media Technologies: Essays on Communication, Materiality, and Society*, 241–250. Cambridge, MA: MIT Press.

Livingstone, S. (2008) "Taking Risky Opportunities in Youthful Content Creation: Teenager's Use of Social Networking Sites for Intimacy, Privacy and Self-Expression." *New Media & Society* (10)3: 393–410.

Lomborg, S., & Bechmann, A. (2014) "Using APIs for Data Collection on Social Media." *Information and Society* 30: 256–265. doi:10.1080/01972243.2014.91527.

Lovink, G. (2011) *Networks without a Cause.* Cambridge: Polity Press.

Lovink, G. (2016) *Social Media Abyss, Critical Internet Culture and the Force of Negation.* Cambridge: Polity Press.

Lynch, M. (2013) "Privacy and the Threat to the Self." *New York Times.* Available at: http://opinionator.blogs.nytimes.com/2013/06/22/privacy-and-the-threat-to-the-self/?_php=true&_type=blogs&_r=0 (accessed December 14, 2013).

Lyon, D. (2014) "Surveillance, Snowden, and Big Data: Capacities, Consequences, Critique." *Big Data and Society* 1, 1–13. doi:10.1177/2053951714541861.

MacAskill, E., Borger, J., Hopkins, N., Davies, N., & Ball, J. (2013). "GCHQ Taps Fibre-Optic Cables for Secret Access to World's Communications." *The Guardian*, June 27 [Online]. Available at: http://www.theguardian.com/uk/2013/jun/21/gchq-cables-secret-world-communications-nsa (accessed August 19, 2013).

Mai, J. (2016) "Big Data Privacy: The Datafication of Personal Information." *The Information Society* 32(3): 192–199. doi:http://dx.doi.org/10.1080/01972243.2016.1153010.

Manovich, L. (2011) "Trending: The Promises and the Challenges of Big Social Data." Available at: http://manovich.net/index.php/projects/trending-the-promises-and-the-challenges-of-big-social-data (accessed April 17, 2013).

Marcus, G. E. (1995) "Ethnography in/of the World System: The Emergence of Multi-Sited Ethnography." *Annual Review of Anthropology* 24: 95–117. doi:10.1146/annurev.an.24.100195.000523.

Marwick, A. (2013) "Online Identity." In J. Hartley, J. Burgess, & A. Burns (eds.), *The Companion to New Media Dynamics.* London: Blackwell, pp. 365–374.

Marwick, A., & boyd, d. (2011) "I Tweet Honestly, I Tweet Passionately: Twitter Users, Context Collapse and the Imagined Audience." *New Media & Society* 13(1): 114–133.

Marwick, A., & boyd, d. (2014) "Networked Privacy: How Teenagers Negotiate Context in Social Media." *New Media & Society*. 16(7), 1051–1067. doi:10.1177/1461444814543995.

Mattelart, A., & Mattelart, M. (1998) *Theories of Communication: A Short Introduction*. London: SAGE.

Maynard, M., & Purvis, J. (1994) *Researching Women's Lives from a Feminist Perspective*. London: Taylor & Francis.

Mayer-Schönberger, V., & Ramge, T. (2018) *Reinventing Capitalism in the Age of Big Data*. London: John Murray.

Mazzetti, M., & Schmidt, M. (2013) "Edward Snowden, Ex-C.I.A. Worker, Says He Disclosed U.S. Surveillance." *New York Times*, March 9 [Online]. Available at: http://www.nytimes.com/2013/06/10/us/former-cia-worker-says-he-leaked-surveillance-data.html?_r=0 (accessed May 2, 2014).

McGeveran. W. (2015) "Global Internet Privacy Regulation and the 'Right to Be Forgotten.'" *Oxford Internet Summer Doctoral Programme*, University of Oxford, UK, July 2015.

McLaughlin, C., & Vitak, J. (2012) "Norm Evolution and Violation on Facebook." *New Media & Society* 14(2): 299–315.

McNeill, P., & Chapman, S. (2005) *Research Methods*. New York: Psychology Press.

McStay, A. (2012) "I Consent: An Analysis of the Cookie Directive and Its Implications for UK Behavioral Advertising." *New Media and Society*, 15 (4), 596–611 doi:10.1177/1461444812458434.

McStay, A. (2017) *Privacy and the Media*. London: SAGE.

Mead, G. (1934) *Mind, Self, and Society*. Chicago: University of Chicago Press.

Meyer, J. (2011) *Tracking the Trackers: Self Help Tools*. Available at: http://cyberlaw.stanford.edu/blog/2011/09/tracking-trackers-self-help-tools (accessed September 15, 2013).

Microsoft. (2018) "Cookie Notice." *Hotmail, Outlook, Skype, Bing, Latest News, Photos and Videos*. Available at: http://www.msn.com/en-gb/ (accessed January 23, 2018).

Miller, D. (2011) *Tales from Facebook*. London: Polity.

MIT. (2013) "How to Burst the 'Filter Bubble' That Protects Us from Opposing Views." *MIT Technology Review*. Available at: https://www.technologyreview.com/s/522111/how-to-burst-the-filter-bubble-that-protects-us-from-opposing-views/ (accessed September 7, 2018).

Mladenov, T., Owens, J., & Cribb, A. (2015) "Personalisation in Disability Services and Healthcare: A Critical Comparative Analysis." *Critical Society Policy* 35: 307–326. doi:10.1177/0261018315587071.

Mosseri, A. (2016) "Building a Better News Feed for You." *Facebook Newsroom*. Available at: https://newsroom.fb.com/news/2016/06/building-a-better-news-feed-for-you/ (accessed September 2, 2018).

Mowlabocus, S. (2014) "The 'Mastery' of the Swipe: Smartphones and Precarity in a Culture of Narcissism." Paper presented to the Internet Research Conference 16: The 16th Annual Meeting of the Association of Internet Researchers, Phoenix, AZ, October 21–24, 2014.

MSN Cookie Notice. (2018) *Homepage.* Available at: http://www.msn.com/en-gb/ (accessed March 11, 2018).

Murdock, G., & Golding, P. (2005) "Culture, Communication and Political Economy." In J. Curran & M. Gurevitch (eds.), *Mass Media and Society* (4th ed.). London: Hodder, pp. 60–83.

Myers, K., Berry, P., Blythe, J., Conley, K., Gervasio, M., McGuinness, D., Morley, D., Pfeffer, A., Pollack, M., & Tambe, M. (2007) "An Intelligent Personal Assistant for Task and Time Management." *AI Magazine* 28(2): 47–61.

Nakamura, L. (2002) *Cybertypes: Race, Ethnicity and Identity on the Internet.* London: Routledge.

Neff, G., & Nagy, P. (2016) "Automation, Algorithms, and Politics| Talking to Bots: Symbiotic Agency and the Case of Tay." *International Journal of Communication* 10: 4915–4931.

Negroponte, N. (1996) *Being Digital.* New York: Coronet.

NetMarketShare. (2018) "Search Engine Market Share." *NetMarketShare,* Available at: https://netmarketshare.com/search-engine-market-share.aspx (accessed September 7, 2018).

Netvibes. (2016) *Decision-Making Dashboard.* Available at: https://www.netvibes.com/en (accessed May 2, 2016).

NHS. (2017) *Personalised Care.* Available at: http://www.nhs.uk/Conditions/social-care-and-support-guide/Pages/personalisation.aspx) (accessed June 23, 2017)

Nikiforakis, N., Kapravelos, A., Joosen, W., Kruegel, C., Piessens, F., & Vigna, G. (2013) "Cookieless Monster: Exploring the Ecosystem of Web-Based Device Fingerprinting." *IEEE Symposium on Security and Privacy.* Berkeley, CA, pp. 541–555. doi:10.1109/SP.2013.43.

Nissenbaum, A., & Shifman, L. (2015) "Internet Memes as Contested Cultural Capital: The Case of 4chan's /b/ Board." *New Media & Society* 19(4): 483–501. https://doi.org/10.1177/1461444815609313.

Nissenbaum, H. (2010) *Privacy in Context: Technology, Policy, and the Integrity of Social Life.* Stanford, CA: Stanford Law Press.

Noble, S. U. (2018) *Algorithms of Oppression: How Search Engines Reinforce Racism.* New York: New York University Press.

Noys, B. (2015) "Drone Metaphysics." *Culture Machine* 16.

Oath. (2019) "Oath Privacy Policy." *Oath: A Verizon Company.* Available at: https://privacy.aol.com/legacy/privacy-policy.1.html (accessed June 25, 2019).

O'Brien, J. (1999) "Writing in the Body: Gender (Re)production in Online Interaction." In M. A Smith & P. Kollock (eds.), *Communities in Cyberspace.* London: Routledge, pp. 76–106.

OED. (2018) "Chav." *OED Online.* Available at: http://www.oed.com/view/Entry/264253 (accessed February 12, 2016).

Ofcom. (2015) *Online Data and Privacy: Final Report.* Available at: http://stakeholders.ofcom.org.uk/binaries/internet/personal-data-and-privacy/Personal_Data_and_Privacy.pdf (accessed January 14, 2016).

Ofcom. (2019) *Adtech Market Research Report.* Available at: https://www.ofcom.org.uk/__data/assets/pdf_file/0023/141683/ico-adtech-research.pdf (accessed May 11, 2019).

O'Neil, C. (2016) *Weapons of Math Destruction: How Big Data Increases Inequality and Threatens Democracy.* London: Penguin.

Open Banking Europe. (2018) "Building a Digital Future Together." *Open Banking Europe.* Available at: https://www.openbankingeurope.eu/ (accessed September 9, 2018).

Open Rights Group. (2013) *Digital Privacy.* Available at: http://www.openrightsgroup.org/campaigns/digitalprivacy (accessed November 2, 2016).

Optimizely. (2018) "Being Personal Is No Longer Optional." *Optimizely.* Available at: https://www.optimizely.com/products/personalization/ (accessed September 6, 2018).

Oudshoorn, N., Rommes, E., & Stienstra, M. (2004) "Configuring the User as Everybody: Gender and Design Cultures in Information and Communication Technologies." *Science, Technology and Human Values* 29(1): 30–63.

Outbrain. (2014) *Reach New Audiences.* Available at: http://www.outbrain.com/amplify (accessed November 16, 2014).

Pariser, E. (2011) *The Filter Bubble: What the Internet Is Hiding from You.* London: Penguin.

Pariser, E. (2011a) *Eli Pariser: Beware Online Filter Bubbles.* Available at: http://www.ted.com/talks/eli_pariser_beware_online_filter_bubbles.html (accessed September 15, 2011).

Pasquale, F. (2015) "The Algorithmic Self." *The Hedgehog Review* 17(1).

Pazzani, M. J., & Billsus, D. (2007) "Content-Based Recommendation Systems." In Brusilovski, P., Kobsa, A., Nejdl. W. *The Adaptive Web.* Berlin: Springer, pp. 325–341.

Peacock, S. (2014) "How Web Tracking Changes User Agency in the Age of Big Data: The Used User." *Big Data & Society*, 1(2). doi:http://dx.doi.org/10.1177/2053951714564228.

Personalising Education. (2017) *Homepage.* Available at: http://www.personalisingeducation.org/ (accessed June 23, 2017).

Peters, B. (2012) "The Big Data Gold Rush." *Forbes.* Available at: https://www.forbes.com/sites/bradpeters/2012/06/21/the-big-data-gold-rush/#734fcaaab247 (accessed September 8, 2018).

Pfeffer, A., Pollack, M., & Tambe, M. (2007) "An Intelligent Personal Assistant for Task and Time Management." *AI Magazine* 28(2): 47–61.

Ponsot, E. (2017) "A Complete Guide to Seeing the News beyond Your Cozy Filter Bubble." *Quartz.* Available at: https://qz.com/896000/a-complete-guide-to-seeing-beyond-your-cozy-filter-bubble/ (accessed September 7, 2018).

Poster, M. (2006) *Information Please: Culture and Politics in the Age of Digital Machines.* Durham, NC: Duke University Press.

Preston, A. (2014) "The Death of Privacy." *The Guardian.* Available at: https://www.theguardian.com/world/2014/aug/03/internet-death-privacy-google-facebook-alex-preston (accessed September 6, 2018).

Prezzie Box. (2017) *Personalised Gifts.* Available at: http://www.prezzybox.com/personalised-gifts.aspx (accessed June 23, 2017).

Protalinski, E. (2018) "Over 90% of Facebook's Advertising Revenue Now Comes from Mobile." *Venture Beat.* Available at: https://venturebeat.com/2018/04/25/

over-90-of-facebooks-advertising-revenue-now-comes-from-mobile/ (accessed September 2, 2018).

Qq (2019) *Tencent Privacy Protection Platform*. Available at: https://privacy.qq.com/ (accessed November 5, 2019).

Quantcast. (2013) *Quantcast Measure*. Available at: https://www.quantcast.com/measure/(accessed June 5, 2013).

Rainie, L., &, Wellman, B. (2014) *Networked: The New Social Operating System*. Cambridge, MA: MIT Press.

Reddit. (2017) *Reddit Inc., Privacy Policy*. Available at: https://www.reddit.com/help/privacypolicy (accessed June 28, 2017).

Rheingold, H. (1996). *The Virtual Community: Homesteading on the Electronic Frontier*. Cambridge, MA: MIT Press.

Ringrose, J., & Walkerdine, V. (2008) "Regulating the Abject: The TV Make-over as Site of Neo-liberal Reinvention toward Bourgeois Femininity." *Feminist Media Studies* 8: 227–246.

Roche. (2017) *Personalised Healthcare at Roche*. Available at http://www.roche.com/about/priorities/personalised_healthcare.htm (accessed June 23, 2017).

Roberts, D., & Parks, M. (1999) "The Social Geography of Gender-Switching in Virtual Environments of the Internet." In E. Green & A. Adam (eds.), *Virtual Gender: Technology, Consumption and Identity*. London: Routledge, pp. 265–286.

Rose, N. (1991) *Governing the Soul: The Shaping of the Private Self*. London: Free Association Books.

Roush, W. (2006) *What Comes After Web 2.0?*. Available at: https://www.technologyreview.com/s/406937/what-comes-after-web-20/. (accessed November 5, 2019).

Ruppert, E., Law, J., & Savage, M. (2013) "Reassembling Social Science Methods: The Challenge of Digital Devices." *Theory, Culture & Society* 30(4): 22–46.

Said, A. & Bellogín (2018) "Coherence and Inconsistencies in Rating Behavior: Estimating the Magic Barrier of Recommender Systems." *User Modeling and User-Adapted Interaction* 28(2): 97–125.

Sartre, J.-P. (1957) *The Transcendence of the Ego: An Existentialist Theory of Consciousness*. New York: Noonday.

Sauter, T. (2013): "'What's on Your Mind?' Writing on Facebook as a Tool for Self-Formation." *New Media & Society* 16: 1–17.

Scannell, P. (2005) "The Meaning of Broadcasting in the Digital Era." In G. Lowe & P. Jauert (eds.), *Cultural Dilemmas in Public Service Broadcasting*. Goteborg: Nordicom, pp. 129–141.

Seeman, M. (2015) *Digital Tailspin: Ten Rules for the Internet after Snowden*. Amsterdam: Institute of Network Culture.

Simonite, T. (2013) "A Popular Ad Block That Also Helps the Ad Industry," *MIT Technology Review*. Available at: http://www.technologyreview.com/news/516156/a-popular-ad-blocker-also-helps-the-ad-industry/ (accessed September 17, 2013).

Silverstone, R. (1994) *Television and Everyday Life*. London: Routledge.

Skeggs, B. (2004) *Class, Self, Culture*. London: Routledge.

Skeggs, B. (2011) "Imagining Personhood Differently: Person Value and Autonomist Working-Class Value Practices." *The Sociological Review* 59(18): 496.

Skeggs, B. (2017) *You Are Being Tracked, Valued and Sold: An Analysis of Digital Inequalities*. Available at: http://www.lse.ac.uk/Events/Events-Assets/PDF/2017/2017-MT03/20170926-Bev-Skeggs-PPT.pdf (accessed September 2, 2018).

Skeggs, B., Thumim, N., & Wood, H (2008) "'Oh goodness, I am watching reality TV': How Methods Make Class in Audience Research." *European Journal of Cultural Studies* 11: 5–24. doi:10.1177/1367549407084961.

Skeggs, B., & Yuill, S. (2016) "The Methodology of a Multi-Model Project Examining How Facebook Infrastructures Social Relations." *Information, Communication & Society* 19(10): 1356–1372.

Skrubbeltrang, M. M, Grunnet, J., & Tarp, N. T. (2017) "#RIPINSTAGRAM: Examining User's Counter-Narratives Opposing the Introduction of Algorithmic Personalization on Instagram." *First Monday*, 4(3). doi:https:// doi.org/10.5210/fm.v22i4.7574.

Smithers, T. (1997) "Autonomy in robots and other agents." *Brain and Cognition*, 34: 88–106.

Smith-Shomade, B. E. (2004) "Narrowcasting in the New World Information Order: A Space for the Audience?" *Television & New Media* 5(1): 68–81.

Solon, O. (2018) "Sheryl Sandberg: Facebook Business Chief Leans out of Spotlight in Scandal." *The Guardian*, March 28 [Online]. Available at: https://www.theguardian.com/technology/2018/mar/29/sheryl-sandberg-facebook-cambridge-analytica (accessed September 10, 2018).

Spotify. (2018) "Spotify Privacy Policy." *Spotify*. Available at: https://www.spotify.com/uk/legal/privacy-policy/ (accessed September 2, 2018).

Spotify Community. (2013) *Forced to Connect to Spotify*. Available from: https://community.spotify.com/t5/Help-Accounts-and-Subscriptions/Forced-to-connect-to-facebook/td-p/522602 (accessed June 23, 2014).

Stalder, F., & Mayer, C. (2009) "The Second Index: Search Engines, Personalization and Surveillance." In R. Becker & F. Stalder (eds.), *Deep Search: The Politics of Search beyond Google*. Innsbruck: StudienVerlag, pp. 98–116.

Stanley, L., & Wise, S. (1990) "Method, Methodology and Epistemology in Feminist Research Processes." In L. Stanley (ed.), *Feminist Praxis: Research, Theory and Epistemology in Feminist Sociology*. London: Routledge.

Statista, (2019). "Number of Spotify monthly active users (MAUs) worldwide from 1st quarter 2015 to 2nd quarter 2019 (in millions)." *Statista*. Available at: https://www.statista.com/statistics/367739/spotify-global-mau/ (accessed November 3, 2019).

Statistic Brain. (2014) *Facebook Statistics*. Available at: http://www.statisticbrain.com/facebook-statistics (accessed June 12, 2014).

Stone, A. (1995) *The War of Desire and Technology at the Close of the Mechanical Age*. Cambridge, MA: MIT Press.

Sundar, S. S., & Marathe, S. S. (2010) "Personalization versus Customization: The Importance of Agency, Privacy and Power Usage." *Human Communication Research* 36: 298–322.

Szulc, L. (2018) "Profiles, Identities, Data: Making Abundant and Anchored Selves in Platform Society." *Communication Theory* 29(3): 257–276.

TechCrunch. (2014) "Google Now Has 1B Active Monthly Android Users." *Techcrunch*, June 25 [Online] Available at: http://techcrunch.com/2014/06/25/google-now-has-1b-active-android-users/ (accessed June 5, 2014).

Terranova, T. (2000) "Free Labour: Producing Culture in the Digital Economy." *Social Text* 18(2): 33–58.

Thumin, N. (2015) *Self-Representation and Digital Culture*. London: Palgrave.

TMall. (2018) "Privacy Policy." *TMall.* Available at: https://rule.tmall.com/tdetail-6684.htm?spm=a225s.11047550.a2226n1.31.79d1480dS5sgZb&tag=self (accessed September 7, 2018).

Trip Advisor. (2017) *Privacy Policy.* Available at: https://tripadvisor.mediaroom.com/UK-privacy-policy (accessed June 23, 2017).

Truste. (2016) *Privacy Compliance, Risk Management and Trust.* Available at: https://www.truste.com/ (accessed March 23, 2016).

Turkle, S. (1997) *Life on the Screen: Identity in the Age of the Internet*. New York: Simon & Schuster.

Turkle, S. (2011) *Alone Together: Why We Expect More from Teachnology and Less from Eachother*. New York: Basic Books.

Turner, F. (2014) "The World Outside and the Pictures Inside Our Networks." In T. Gillespie, P. Boczkowski, & K. Foot (eds.), *Media Technologies: Essays on Communication, Materiality, and Society*. Cambridge, MA: MIT Press.

Turow, J. (2012) *The Daily You: How the Advertising Industry Is Defining Your Identity and Your World.* New Haven, CT: Yale University Press.

Turow, J., Hennesy, M., & Draper, N. (2015) "The Tradeoff Fallacy: How Marketers Are Misrepresenting American Consumers and Opening Them Up to Exploitation." Report from The Anneberg School for Communication, California.

Twitter. (2015) *About Tailored Suggestions*. Available at: https://support.twitter.com/articles/20169421-about-tailored-suggestions# (accessed May 4, 2015).

Twitter. (2017) *Privacy Policy*. Available at: https://twitter.com/en/privacy (accessed June 28, 2017).

Twitterclips. (2018) "Alexa, What's 5 – 3? | Alexa Doing Homework for This Little Boy." *YouTube.* Available at: https://www.youtube.com/watch?v=9yq3zlJzdls (accessed June 25, 2019).

Vaidhynathan, S. (2011) *The Googlization of Everything: (And Why We Should Worry).* Berkeley: University of California Press.

Van Couvering, E. (2007) "Is Relevance Relevant? Market, Science, and War: Discourses of Search Engine Quality." *Journal of Computer Mediated Communication* 12: 866–887.

Van Dijck, J. (2009) "Users like You? Theorizing Agency in User-Generated Content." *Media, Culture and Society* 31(1): 41–58.

Van Dijck, J. (2013) "'You Have One Identity': Performing the Self on Facebook and LinkedIn." *Media, Culture and Society* 35(2): 199–215.

Vesanen, J., & Raulas, M. (2006) "Building Bridges for Personalization: A Process Model for Marketing." *Journal of Interactive Marketing* 20: 5–20. doi:10.1002/dir.20052.

Vicente-López, E., de Campos, L. M., Fernández-Luna, J. M., Huete, J. F., Tagua-Jiménez A., & Tur-Vigil, C. (2015) "An automatic methodology to evaluate personalized information retrieval systems." *Journal of User Modelling and User-Adaptive Interaction* 25: 1–37. doi:http://dx.doi.org/10.1007/s11257-014-9148-9.

Vidler, A. (1992) *The Architectural Uncanny: Essays in the Modern Unhomely*. Cambridge, MA: MIT Press.

Wang, J., Zhang, W., & Yuan, S. (2017). "Display Advertising with Real-Time Bidding (RTB) and Behavioural Targeting." *ArXiv.org*.

Walker, J. (2014) *Seeing Ourselves through Technology: How We Use Selfies, Blogs and Wearable Devices to See and Shape Ourselves*. Basingstoke: Palgrave Macmillan.

Warren, C. (2001) "Qualitative Interviewing." In J. F. Gubrium & J. A. Holstein (eds.), *Handbook of Interview Research*. Thousand Oaks, CA: SAGE, pp. 83–103.

Watkins, D. (2018) "Strategy Analytics: Explosive Growth for Smart Speakers Continues as Global Sales Top 18 Million Units in Q4 2017." *Strategy Analytics*. Available at: https://www.strategyanalytics.com/strategy-analytics/news/strategy-analytics-press-releases/strategy-analytics-press-release/2018/02/27/strategy-analytics-explosive-growth-for-smart-speakers-continues-as-global-sales-top-18-million-units-in-q4-2017#.WpWH2ZPwb-Y (accessed September 7, 2018).

Watson, J. (2013) "Facebook Wants to Listen to Your Phone Calls." *Infowars*. Available at: http://www.infowars.com/facebook-wants-to-listen-to-your-phone-calls/ (accessed February 15, 2014).

Weber, M. ([1947] 2012) *The Theory of Social and Economic Organization*. London: Martino.

Westling, C. E. I. (2020) *Immersion and Participation in Punchdrunk's Theatrical Worlds*. London: Bloomsbury.

Willson, M. & Leaver, T. (2015) "Zynga's FarmVille, Social Games, and the Ethics of Big Data Mining." *Communication Research and Practice* 1(2): 147–158.

Whittaker, Z. (2016) "NSA Is So Overwhelmed with Data, It's No Longer Effective, Says Whistleblower." *ZDNet*. Available at: https://www.zdnet.com/article/nsa-whistleblower-overwhelmed-with data-ineffective/ (accessed September 6, 2018).

Wyatt, S. (2005) "Non-users Also Matter: The Construction of Non-users and Users of the Internet." In N. Oushoorden & T. Pinch (eds.), *How Users Matter: The Co-construction of Users and Technology*. Cambridge, MA: MIT Press.

Yahoo! (2017) *Privacy Policy*. Available at: https://policies.yahoo.com/us/en/yahoo/privacy/index.htm (accessed June 28, 2017).

Your Design. (2017) *Create Customised Gifts*. Available at: https://www.yourdesign.co.uk/create.html (accessed June 23, 2017).

YouTube. (2015) *Advertise*. Available at: https://www.youtube.com/yt/advertise/en-GB/ (accessed May 12, 2015).

YouTube. (2016) *Homepage*. Available at: https://www.youtube.com/?gl=GB (accessed April 29, 2016).

ZDNet. (2011) Spotify Defends New Facebook Requirement. Available at: http://www.zdnet.com/article/spotify-defends-new-facebook-requirement/ (accessed August 15, 2014).

Zhu, Q., Zhou, M., Xia, Y., & Zhu. Q, (2014) "An Efficient Non-Negative Matrix-Factorization-Based Approach to Collaborative Filtering for Recommender Systems." *IEEE Transactions on Industrial Informatics 10*(2): 1273–1284. http://doi.org/10.1109/TII.2014.2308433

Zuckerberg, M. (2011) *F82011 Keynote*. Available at: http://www.youtube.com/watch?v=ZLxlJbwxukA (accessed January 29, 2013).

Zuckerberg, M., Sukhar, I., Archibong, I., & Liu, D. (2014) *F8 2014 Keynote*. https://www.youtube.com/watch?v=0onciIB-ZJA (accessed September 18, 2014).

Zylinska, J. (2001) *Spiders, Cyborgs and Being Scared: The Feminine and the Sublime*. Manchester: Manchester University Press.

INDEX

For the benefit of digital users, indexed terms that span two pages (e.g., 52–53) may, on occasion, appear on only one of those pages.

www.ingramcontent.com/pod-product-compliance
Ingram Content Group UK Ltd.
Pitfield, Milton Keynes, MK11 3LW, UK
UKHW021135160926
475595UK00020B/56

9 780190 905088